NIST Special Publication 800-21

[Second Edition]

NIST
National Institute of Standards and Technology
Technology Administration
U.S. Department of Commerce

Guideline for Implementing Cryptography In the Federal Government

**Elaine B. Barker, William C. Barker,
Annabelle Lee**

INFORMATION SECURITY

Computer Security Division
Information Technology Laboratory
National Institute of Standards and Technology
Gaithersburg, MD 20899-8930

December 2005

U.S. Department of Commerce

Reports on Computer Systems Technology

The Information Technology Laboratory (ITL) at the National Institute of Standards and Technology (NIST) promotes the U.S. economy and public welfare by providing technical leadership for the Nation's measurement and standards infrastructure. ITL develops tests, test methods, reference data, proof of concept implementations, and technical analyses to advance the development and productive use of information technology. ITL's responsibilities include the development of management, administrative, technical, and physical standards and guidelines for the cost-effective security and privacy of non-national security-related information in Federal information systems. This special publication 800-series reports on ITL's research, guidelines, and outreach efforts in information system security, and its collaborative activities with industry, government, and academic organizations.

Authority

The National Institute of Standards and Technology (NIST) has developed this document in furtherance of its statutory responsibilities under the Federal Information Security Management Act (FISMA) of 2002, Public Law 107-347.

NIST is responsible for developing standards and guidelines, including minimum requirements, for providing adequate information security for all agency operations and assets, but such standards and guidelines shall not apply to national security systems. This guideline is consistent with the requirements of the Office of Management and Budget (OMB) Circular A-130, Section 8b(3), Securing Agency Information Systems, as analyzed in A-130, Appendix IV: Analysis of Key Sections. Supplemental information is provided A-130, Appendix III.

This guideline has been prepared for use by Federal agencies. It may be used by nongovernmental organizations on a voluntary basis and is not subject to copyright. (Attribution would be appreciated by NIST.)

Nothing in this document should be taken to contradict standards and guidelines made mandatory and binding on Federal agencies by the Secretary of Commerce under statutory authority. Nor should these guidelines be interpreted as altering or superseding the existing authorities of the Secretary of Commerce, Director of the OMB, or any other Federal official.

National Institute of Standards and Technology, Special Publication 800-21

Natl. Inst. Stand. Technol. Spec. Publ. 800-21 [2nd Edition], (*December 2005*)

Acknowledgments

The authors wish to thank their colleagues who reviewed drafts of this document and contributed to its development. The authors also gratefully acknowledge and appreciate the many comments from the public and private sectors whose thoughtful and constructive comments improved the quality and usefulness of this publication.

PREFACE

This Second Edition of NIST Special Publication (SP) 800-21, updates and replaces the November 1999 edition of *Guideline for Implementing Cryptography in the Federal Government*. Many of the references and cryptographic techniques contained in the first edition of NIST SP 800-21 have been amended, rescinded, or superseded since its publication. The current revision offers new tools and techniques.

NIST SP 800-21 [Second Edition] is intended to provide a structured, yet flexible set of guidelines for selecting, specifying, employing, and evaluating cryptographic protection mechanisms in Federal information systems—and thus, makes a significant contribution toward satisfying the security requirements of the Federal Information Security Management Act (FISMA) of 2002, Public Law 107-347. The current version also reflects the elimination of the waiver process by the Federal Information Security Management Act (FISMA) of 2002. Under current law, NIST standards and recommendations are binding for Federal systems that are not designated *national security systems*.

GUIDELINE FOR IMPLEMENTING CRYPTOGRAPHY IN THE FEDERAL GOVERNMENT

1. INTRODUCTION 1
   1.1 Purpose.. 1
   1.2 Audience 2
   1.3 Scope... 2
   1.4 Content 3
   1.5 Uses of Cryptography 4

2. STANDARDS AND GUIDELINES... 6
   2.1 Benefits of Standards.. 7
   2.2 Federal Information Processing Standards (FIPS) and Special Publications (SPs) 8
       2.2.1 Use of FIPS and SPs.. 8
       2.2.2 FIPS Waivers 9
   2.3 Other Standards Organizations... 9
       2.3.1 International Organization for Standardization (ISO) 9
       2.3.2 American National Standards Institute (ANSI).................................. 10
       2.3.3 Institute of Electrical and Electronics Engineers (IEEE) 10
       2.3.4 Internet Engineering Task Force (IETF) 10

3. CRYPTOGRAPHIC METHODS.. 12
   3.1 Overview of Cryptography... 12
   3.2 Hash Functions 13
   3.3 Symmetric Key Algorithms 14
       3.3.1 Encryption and Decryption 14
            3.3.1.1 Data Encryption Standard (DES) 15
            3.3.1.2 Triple Data Encryption Algorithm (TDEA).......................... 15
            3.3.1.3 Advanced Encryption Standard (AES) 16
            3.3.1.4 Encryption Modes of Operation... 16
       3.3.2 Message Authentication Code... 16

           3.3.2.1 MAC Based on a Block Cipher Algorithm 17
           3.3.2.2 MACs Based on Hash Functions 18
      3.3.3 Key Establishment 18
  3.4 Asymmetric Key Algorithms 18
      3.4.1 Digital Signatures and the Digital Signature Standard (DSS) 19
      3.4.2 Key Establishment 21
  3.5 Random Number Generation 22
  3.6 Key Management 22
  3.7 Public Key Infrastructure (PKI) 24
      3.7.1 Security Requirements for PKI Components 26
      3.7.2 PKI Architectures 26
      3.7.3 Security Policies of Other CAs and the Network 27
      3.7.4 Federal Bridge Certification Authority 28

4. GENERAL IMPLEMENTATION ISSUES 29
  4.1 Hardware vs. Software Solutions 29
  4.2 Asymmetric vs. Symmetric Cryptography 30
  4.3 Key Management 31

5. ASSESSMENTS 34
  5.1 Cryptographic Module Validation Program (CMVP) 35
      5.1.1 Background 36
      5.1.2 FIPS 140-2 Requirements 38
      5.1.3 Pre-Validation List 39
      5.1.4 Validated Modules List 40
      5.1.5 Effective Use of FIPS 140-2 41
  5.2 National Voluntary Laboratory Accreditation Program (NVLAP) 42
  5.3 Industry and Standards Organizations 42
      5.3.1 National Information Assurance Partnership (NIAP) 42
      5.3.2 Certification and Accreditation 44

6. SELECTING CRYPTOGRAPHY - THE PROCESS 46

6.1 Phase 1: Initiation 50
    6.1.1 Business Partner Engagement and Document Enterprise Architecture 50
    6.1.2 Identify/Specify Applicable Policies and Laws 50
    6.1.3 Develop C, I, and A Objectives 52
    6.1.4 Information and Information System Security Categorization and Procurement Specification Development 52
    6.1.5 Cryptographic Method Example 53
    6.1.6 Preliminary Risk Assessment 57

6.2 Phase 2: Acquisition/Development 59
    6.2.1 Selecting Cryptographic Controls 59

6.3 Phase 3: Implementation/Assessment 67

6.4 Phase 4: Operations and Maintenance 69

6.5 Phase 5: Sunset (Disposition) 70

APPENDIX A: ACRONYMS 71
APPENDIX B: TERMS AND DEFINITIONS 74
APPENDIX C: REFERENCE LIST 80
APPENDIX D: INFORMATION SECURITY LAWS AND REGULATIONS 85
APPENDIX E: APPLICABLE FIPS AND SPECIAL PUBLICATIONS 87

CHAPTER 1

INTRODUCTION

1.1 Purpose

Today's information technology security environment consists of highly interactive and powerful computing devices and interconnected systems of systems across global networks where Federal agencies routinely interact with industry, private citizens, state and local governments, and the governments of other nations. Consequently, both private and public sectors depend upon information systems to perform essential and mission-critical functions. In this environment of increasingly open and interconnected systems and networks, network and data security are essential for the optimum use of this information technology. For example, systems that carry out electronic financial transactions and electronic- commerce (e-commerce) must protect against unauthorized access to confidential records and the unauthorized modification of data.

Cryptography should be considered for data that is sensitive, has a high value, or is vulnerable to unauthorized disclosure or undetected modification during transmission or while in storage[1]. Cryptographic methods provide important functionality to protect against intentional and accidental compromise and alteration of data. Some cryptographic mechanisms support confidentiality during communications by encrypting the communication prior to transmission and decrypting it at receipt. These methods also provide file/data confidentiality by encrypting the data prior to placement on a storage medium and decrypting it after retrieval from the storage medium. Other cryptographic mechanisms, such as message authentication codes and digital signatures, provide data content integrity and source authentication services. That is, the cryptographic mechanisms permit the user to determine that the entity claiming to be the source of data really is the source and to determine whether information has been modified since it was last authenticated or "signed" by its source.

The purpose of this document is to provide guidance to Federal agencies on how to select cryptographic controls for protecting Sensitive Unclassified[2] information. This document focuses on:

---

[1] FIPS 199, Standards for Security Categorization of Federal Information and Information Systems, provides a standard for categorizing information and information systems, based on the impact to the mission if the confidentiality, integrity or availability of the information was compromised. NIST SP 800-53, Recommended Security Controls for Federal Information Systems, provides guidance on the minimum security controls for each FIPS 199 category.

[2] Hereafter referred to as sensitive information. In the Federal Information Security Management Act (FISMA) of 2002, Congress assigned responsibility to the National Institute of Standards and Technology (NIST) for the preparation of standards and guidelines for the security of sensitive *Federal* systems. Excluded are classified and sensitive national security-related systems.

- Federal standards documented in Federal Information Processing Standards (FIPS) Publications,
- NIST Recommendations and guidelines documented in NIST Special Publications (SPs), and
- Cryptographic modules and algorithms that are validated against these specifications.

However, to provide additional information, products of other standards organizations, (e.g., American National Standards Institute (ANSI) and International Organization for Standardization (ISO)) are briefly discussed.

1.2 Audience

This document is intended for Federal employees who are responsible for designing systems, and procuring, installing, and operating security products to meet identified security requirements. This document may be used by:

- A manager responsible for evaluating an existing system and determining whether cryptographic methods are necessary,
- Program managers responsible for selecting and integrating cryptographic mechanisms into a system,
- A technical specialist requested to select one or more cryptographic methods/techniques to meet a specified requirement, or
- A procurement specialist developing a solicitation for a system or network that will require cryptographic methods to perform security functionality.

The goal is to provide these individuals with sufficient information to allow them to make informed decisions about the cryptographic methods that will meet their specific needs to protect the confidentiality, authentication, and integrity of data that is transmitted and/or stored in a system or network.

This document is not intended to provide information on the Federal procurement process or provide a technical discussion on the mathematics of cryptography and cryptographic algorithms.

1.3 Scope

This document limits its discussion of cryptographic methods to those that conform to Federal standards (FIPS) and NIST recommendations (i.e., NIST SPs). (The majority of the information in this guideline may be useful to both Federal and commercial personnel and applicable to all computer networks and environments.) Both the Federal government and industry use products that meet Federal standards and recommendations, and standards bodies such as ANSI have also adopted these documents.

This guideline provides information on selecting cryptographic controls and implementing the controls in new or existing systems. Specifically, the guideline includes discussions of the following:

1. The process of selecting cryptographic products. This may include one or more of the following:

   a. Perform a *risk assessment* that includes the following:
      - System characterization,
      - Threat identification,
      - Vulnerability identification and likelihood determination, and
      - Potential impact on organizations or individuals.

   b. Identify the *security regulations and policies* that are applicable to the organization and system.

   c. Specify the *cryptographic security requirements*.

   d. Specify the *security controls* that will address the needs identified in items a through c above.

   e. Select the appropriate cryptographic mechanism/product for provision of specified security controls.

2. Implementation issues, including:

   a. Implementation approach,

   b. Life cycle management of cryptographic components,

   c. Training of users, operators, and system engineers,

   d. Selection of cryptographic mechanisms,

   e. Key management,

   f. Backup and restoration of services,

   f. Authentication techniques, and

   g. Assessment – certification, independent verification and validation (IV&V).

1.4   Content

This Guideline is organized into six chapters:

- Chapter 1 includes background information (purpose, audience, and scope) and the advantages of using cryptography.

- Chapter 2 defines the role and use of standards and describes standards organizations that are outside the Federal government.
- Chapter 3 describes the methods that are available for symmetric and asymmetric key cryptography.
- Chapter 4 describes some implementation issues (e.g., key management).
- Chapter 5 discusses assessments, including the Cryptographic Module Validation Program (CMVP), the Common Criteria (CC), and Certification and Accreditation (C&A).
- Chapter 6 describes the process of choosing the types of cryptography to be used and selecting a cryptographic method or methods to fulfill a specific requirement.

There are seven appendices to the guideline:

- Appendix A contains an acronym list.
- Appendix B contains terms and definitions.
- Appendix C contains a reference list of cryptographic standards and guidelines and other cryptography references.
- Appendix D lists applicable laws and regulations.
- Appendix E lists applicable Federal information processing standards, recommendations, and guidelines.

A number of examples are included throughout this guideline. Each example is displayed in a shaded box for ease of viewing.

1.5 Uses of Cryptography

Historically, cryptography was used as a tool to protect secrets. Numerous techniques have been used, including:

- Manual systems (e.g., simple substitution, manual codes),
- Mechanical devices (e.g., the World War II and Korean era M 209 device),
- Electro-mechanical devices (e.g., the World War II Enigma and Purple devices), and
- Modern electronic encryption and authentication mechanisms (e.g., Advanced Encryption Standard (AES), Digital Signature Algorithm (DSA), and Keyed Hash Message Authentication Code (HMAC)).

Modern cryptography uses mathematical techniques to provide security services and relies upon two basic components: an algorithm (or cryptographic methodology) and a cryptographic key, which determines the specifics of algorithm operation.

In general, cryptography is used to provide the security objectives of confidentiality, integrity and availability.

- *Confidentiality* addresses "Preserving authorized restrictions on information access and disclosure, including a means for protecting personal privacy and proprietary information..." [44 U.S.C., Sec. 3542]". A loss of *confidentiality* is the unauthorized disclosure of information.

- *Integrity* addresses "Guarding against improper information modification or destruction, and includes ensuring information non-repudiation and authenticity..." [44 U.S.C., Sec. 3542]. A loss of *integrity* is the unauthorized modification or destruction of information.

    o *Non-repudiation* services provide assurance of the origin of data to both the receiver and a third party. The objective is to provide evidence to counter denials that the sender participated in a specified transaction.

    o An assurance of *authenticity* is provided using *authentication controls*, which protect a communication system against acceptance of a fraudulent transmission or simulation by establishing the validity of the information content and the originator. Authentication controls can also be used to verify an individual's *authorization* to access specific categories of information.

- *Availability* addresses "Ensuring timely and reliable access to and use of information..." [44 U.S.C., SEC. 3542]. A loss of *availability* is the disruption of access to or use of information or an information system.

CHAPTER 2

STANDARDS AND GUIDELINES

Chapter 2 addresses standards and guidelines that apply to the implementation of cryptography in the Federal government.

Public Laws, Presidential Directives and Executive Orders, Office of Management and Budget (OMB) Memoranda, etc. (listed in Appendix D) establish requirements for:

- Executive branch departments and agencies to protect all information processed, transmitted, or stored in Federal automated information systems;

- The development and implementation of information security policies, procedures, and control. The controls shall be sufficient to afford security protections that are commensurate with the risk and magnitude of the harm resulting from the unauthorized disclosure, disruption, modification, or destruction of information;

- Assignment of responsibility for security and the development of system security plans for all general support systems and major applications; and

- Agencies to ensure that their information security plans are practiced throughout the lifecycle of each agency system.

In addition, the directives and memoranda:

- Establish the basis and authority for NIST FIPS and SPs (hereafter referred to as NIST-standards), and

- Identify the use of cryptography as a potentially effective security mechanism.

Some of the standards used to protect sensitive information are issued by NIST as FIPS. Other recommendations and guidelines are issued as NIST SPs. Federal agencies shall comply with all mandatory standards, and the agencies are expected to:

- Support the development of such standards,

- Avoid the creation of different standards for government and the private sector, and

- Use voluntary standards whenever possible.

Technically, NIST has the authority to establish standards only for the Federal government. However, NIST standards have a profound effect on commerce and industry. Since NIST standards are developed using a public review process, industry often requires that products conform to these standards. Also,

NIST has a long history of participation in industry standards groups, including the American National Standards Institute (ANSI), International Organization for Standardization (ISO), Institute of Electrical and Electronics Engineers (IEEE), Internet Engineering Task Force (IETF), and others. In some cases, the Federal government adopts industry standards, and in other cases, industry has adopted NIST standards, recommendations, and guidelines as industry standards.

2.1 Benefits of Standards

Standards are important because they define common practices, methods, and measures/metrics. Therefore, standards increase the reliability and effectiveness of products and ensure that the products are produced with a degree of quality. Standards provide solutions that have been accepted by a wide community and evaluated by experts in relevant areas. By using standards, organizations can reduce costs and protect their investments in technology.

Standards provide the following benefits:

- **Interoperability.** Products developed to a specific standard may be used to provide interoperability with other products that conform to the same standard. For example, by using the same cryptographic encryption algorithm, data that was encrypted using vendor A's product may be decrypted using vendor B's product. The use of a common standards-based cryptographic algorithm is necessary, but may not be sufficient to ensure product interoperability. Other common standards, such as communications protocol standards, may also be necessary.

  By ensuring interoperability among different vendors' equipment, standards permit an organization to select from various available products to find the most cost-effective solution.

- **Security.** Standards may be used to establish a common approved level of security. For example, most agency managers are not cryptographic security experts, and, by using a FIPS-approved or NIST-recommended[3] cryptographic algorithm, a manager knows that the algorithm has been found to be adequate for the protection of sensitive government data and has been subjected to a significant period of public analysis and comment.

- **Quality.** Standards may be used to assure the quality of a product. Standards may:

    o Specify how a feature is to be implemented, e.g., the feature must be implemented in hardware.

    o Require a test to ensure that the product is still functioning correctly.

---

[3] Hereafter, FIPS-approved and NIST-recommended are collaterally referred to as Approved.

- Require specific documentation to assure proper implementation and product change management.

Many NIST standards and recommendations contain associated conformance tests and specify the conformance requirements. The conformance tests may be administered by NIST accredited laboratories and provide validation that the NIST standard or recommendation was correctly implemented in the product.

- **Common Form of Reference.** A NIST standard or recommendation may become a common form of reference to be used in testing/evaluating vendors' products. For example, FIPS 140-2, *Security Requirements for Cryptographic Modules*, contains security and integrity requirements for *any* cryptographic module implementing cryptographic operations. FIPS 140-2 establishes a common form of reference by defining four levels of security for each of eleven security attributes.

- **Cost Savings.** A standard can save money by providing a single commonly accepted specification. Without standards, users may be required to become *experts* in every information technology (IT) product that is being considered for purchase. Also, without standards, products may not interoperate with different products purchased by other users. This will result in a significant waste of money or in the delay of implementing IT.

2.2 Federal Information Processing Standards (FIPS) and Special Publications (SPs)

2.2.1 Use of FIPS and SPs

A FIPS is a *mandatory* standard for the Federal government whenever the type of service provided by that standard is required by a Federal agency. For example, FIPS 197, *Advanced Encryption Standard*, is a specific set of technical security requirements for the Advanced Encryption Standard (AES) algorithm. A FIPS has been adopted via a signature by the Secretary of Commerce (SoC).

A NIST recommendation is similar to a FIPS, but has not been signed by the SoC. For example, NIST SP 800-67, *Recommendation for the Triple Data Encryption Algorithm (TDEA) Block Cipher*, provides a similar set of technical security requirements to that of FIPS 197, except that TDEA is specified, rather than AES.

To continue with these examples, when a Federal agency requires the use of encryption to protect its data, an Approved algorithm shall be used. Since AES and TDEA are currently the only algorithms approved for data encryption, either AES or TDEA shall be used. Whenever AES is to be used, it shall be used as specified in FIPS 197; whenever TDEA is to be used, it shall be used as specified in SP 800-67.

When developing a specification or the criteria for the selection of a cryptographic module/product, FIPS and SPs shall be used, when available. Some guidelines may be used to specify the *functions* that the algorithm will perform (e.g., FIPS 200 or NIST SP 800-53, *Recommended Security Controls for Federal Information Systems*). Other NIST standards specify the operation and use of specific types of algorithms (e.g., AES, DSA) and the level of independent testing required for classes of security environments (e.g., FIPS 140-2).

Appendix E contains a list of FIPS and SPs that apply to the implementation of cryptography in the Federal government.

2.2.2 FIPS Waivers

The Federal Information Security Management Act (FISMA) of 2002 (P.L. 107-347) eliminated previously authorized procedures for waivers from FIPS.

2.3 Other Standards Organizations

NIST develops standards, recommendations, and guidelines that are used by vendors who are developing security products, components, and modules. These products may be purchased and used by Federal government agencies. In addition, there are other groups that develop and promulgate standards. The following organizations are briefly described below: ANSI, IEEE, IETF, and ISO.

2.3.1 International Organization for Standardization (ISO)[4]

ISO is a worldwide federation of national standards bodies from 100 countries. ISO is a non-governmental organization. Its mission is to promote the development of standardization and related activities in the world with a view to facilitating the international exchange of goods and services, and to developing cooperation in the spheres of intellectual, scientific, technological and economic activity. ISO's work results in international agreements that are published as International Standards.

The technical work of ISO is carried out in technical committees, subcommittees and working groups. In these committees, qualified representatives of industry, research institutes, government agencies, consumer bodies, and international organizations from all over the world come together in the resolution of global standardization problems.

---

[4] The information in this section was taken from the ISO web site: www.iso.ch.

2.3.2 American National Standards Institute (ANSI) [5]

The American National Standards Institute (ANSI) is the administrator and coordinator of the United States (U.S.) private sector voluntary standardization system. ANSI is a private, nonprofit membership organization that is supported by a diverse constituency of private and public sector organizations. ANSI does not itself develop American National Standards; rather, it facilitates the development of standards by establishing consensus among qualified groups.

The primary goal of ANSI is the enhancement and global competitiveness of U.S. business. ANSI promotes the use of U.S. standards internationally, advocates U.S. policy and technical positions in international and regional standards organizations, and encourages the adoption of international standards as national standards where these meet the needs of the user community.

Accredited Standards Committee X9 is a financial industry committee of ANSI and is organized into sub-committees and working groups to develop guidance in areas such as security, cryptographic tools, and cryptographic protocols. (See www.x9.org.)

2.3.3 Institute of Electrical and Electronics Engineers (IEEE) [6]

The technical objectives of the IEEE focus on advancing the theory and practice of electrical, electronics and computer engineering, and computer science. The goals of IEEE activities are to: (1) enhance the quality of life for all peoples through improved public awareness of the influence and applications of its technologies and (2) advance the standing of the engineering profession and its members.

IEEE develops and disseminates voluntary, consensus-based industry standards involving leading-edge electro-technology. IEEE supports international standardization and encourages the development of globally acceptable standards.

2.3.4 Internet Engineering Task Force (IETF) [7]

The IETF is a large, open international community of network designers, operators, vendors, and researchers concerned with the evolution of the Internet architecture and the smooth operation of the Internet. The actual technical work of the IETF is done in working groups, which are organized by topic into several areas (e.g., routing, transport, security, etc.). A Security Area Directorate and the Security Area Advisory Group has been established to provide help to IETF

---

[5] The information in this section was taken from the ANSI web site: www.ansi.org.

[6] The information in this section was taken from the IEEE web site: www.ieee.org.

[7] The information in this section was taken from the IETF web site: ietf.org.

working groups in providing security in the protocols they design. Working groups are chartered as required to address specific security issues.

CHAPTER 3

CRYPTOGRAPHIC METHODS

This chapter provides a brief overview of cryptography and the various algorithms that are approved for Federal government use.

3.1 Overview of Cryptography

Cryptography is a branch of mathematics that is based on the transformation of data and can be used to provide several security services: confidentiality, data integrity, authentication, authorization and non-repudiation. Cryptography relies upon two basic components: an *algorithm* (or cryptographic methodology) and a *key*. The algorithm is a mathematical function, and the key is a parameter used in the transformation.

A cryptographic algorithm and key are used to apply cryptographic protection to data (e.g., encrypt the data or generate a digital signature) and to remove or check the protection (e.g., decrypt the encrypted data or verify the digital signature). There are three basic types of Approved cryptographic algorithms: cryptographic hash functions, symmetric key algorithms and asymmetric key algorithms:

- Cryptographic hash functions do not require keys (although they can be used in a mode in which keys are used). A hash function is often used as a component of an algorithm to provide a security service. Hash functions are discussed in Section 3.2.

- Symmetric algorithms (often called secret key algorithms) use a single key to both apply the protection and to remove or check the protection. Symmetric key algorithms are discussed in Section 3.3.

- Asymmetric algorithms (often called public key algorithms) use two keys (i.e., a key pair): a public key and a private key that are mathematically related to each other. Asymmetric key algorithms are discussed in Section 3.4.

Random number generators (RNGs) are required for the generation of cryptographic values (e.g., keys). RNGs are discussed in Section 3.5.

In order to use cryptography, cryptographic keys must be "in place", i.e., keys must be established for parties using cryptography. Keys may be established either manually (e.g., via a trusted courier or in a face-to-face meeting) or using an electronic method. However, when an electronic method is used, a manual method of establishing the first key(s) is required. Sections 3.3.3 and 3.4.2 discuss electronic methods for key establishment. Section 3.6 addresses general key management issues, including both manual and electronic methods of key establishment. Section 3.7 discusses Public Key Infrastructures (PKIs), which are used as a method of distributing public keys.

3.2 Hash Functions

A hash function produces a short representation of a longer message. A good hash function is a one-way function: it is easy to compute the hash value from a particular input; however, backing up the process from the hash value back to the input is extremely difficult. With a good hash function, it is also extremely difficult to find two specific inputs that produce the same hash value. Because of these characteristics, hash functions are often used to determine whether or not data has changed.

Many algorithms and schemes that provide a security service use a hash function as a component of the algorithm or scheme. Hash functions are used by:

- Keyed hash message authentication coded algorithms (Section 3.3.2),
- Digital signature algorithms (Section 3.4.1),
- Key derivation functions (e.g., for key agreement) (Section 3.4.2), and
- Random number generators (Section 3.5).

A hash function takes an input of arbitrary length and outputs a fixed length value. Common names for the output of a hash function include hash value and message digest. Figure 1 depicts the use of a hash function. A hash value (H_1) is computed on data (M_1). M_1 and H_1 are then saved or transmitted. At a later time, the correctness of the retrieved or received data is checked by labeling the received data as M_2 (rather than M_1) and computing a new hash value (H_2) on the received value. If the newly computed hash value (H_2) is equal to the retrieved or received hash value (H_1), then it can be assumed that the retrieved or received data (M_2) is the same as the original data (M_1) (i.e., $M_1 = M_2$).

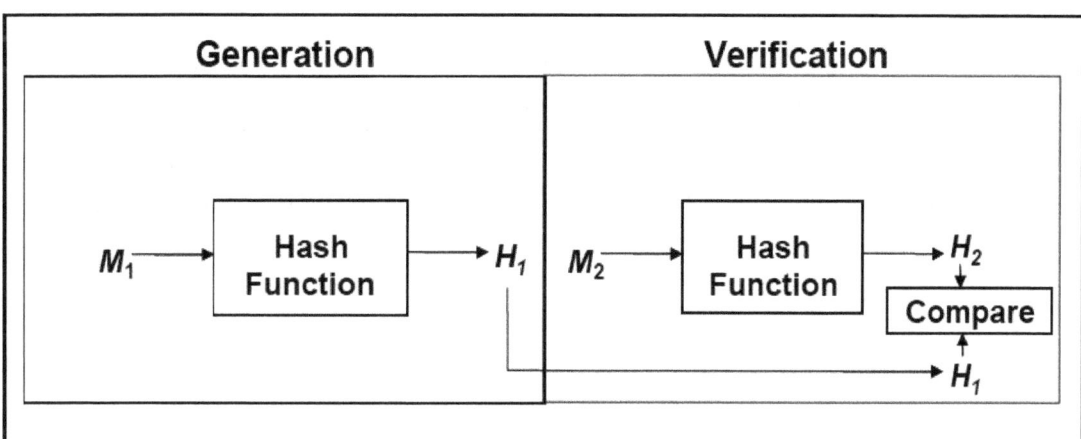

Figure 1: Hash Functions

Five hash functions are approved for Federal Government use and are defined in FIPS 180-2, *Secure Hash Standard*, (http://csrc.nist.gov/publications/fips/fips180-

2/fips180-2change1.pdf). The approved hash functions are SHA-1, SHA-224, SHA-256, SHA-384 and SHA-512. Note that new attacks on SHA-1 have indicated that SHA-1 provides less security than originally thought. The use of SHA-1 is not recommended for the generation of digital signatures in new systems; new systems should use one of the larger hash functions.

3.3 Symmetric Key Algorithms

Symmetric key algorithms (often call secret key algorithms) use a single key to both apply the protection and to remove or check the protection. For example, the key used to encrypt data is also used to decrypt the encrypted data. This key must be kept secret if the data is to retain its cryptographic protection. Symmetric algorithms are used to provide confidentiality via encryption, or an assurance of authenticity or integrity via authentication, or are used during key establishment.

Keys used for one purpose shall not be used for other purposes. (See SP 800-57).

3.3.1 Encryption and Decryption

Encryption is used to provide confidentiality for data. The data to be protected is called plaintext. Encryption transforms the data into ciphertext. Ciphertext can be transformed back into plaintext using decryption. The Approved algorithms for encryption and decryption algorithms are: the Advanced Encryption Standard (AES) and the Triple Data Encryption Algorithm (TDEA). TDEA is based on the Data Encryption Standard (DES), which is no longer approved for Federal Government use except as a component of TDEA. Each of these algorithms operates on blocks (chunks) of data during an encryption or decryption operation. For this reason, these algorithms are commonly referred to as block cipher algorithms.

Plaintext data can be recovered from ciphertext only by using the same key that was used to encrypt the data. Unauthorized recipients of the ciphertext who know the cryptographic algorithm but do not have the correct key should not be able to decrypt the ciphertext. However, anyone who has the key and the cryptographic algorithm can easily decrypt the ciphertext and obtain the original plaintext data.

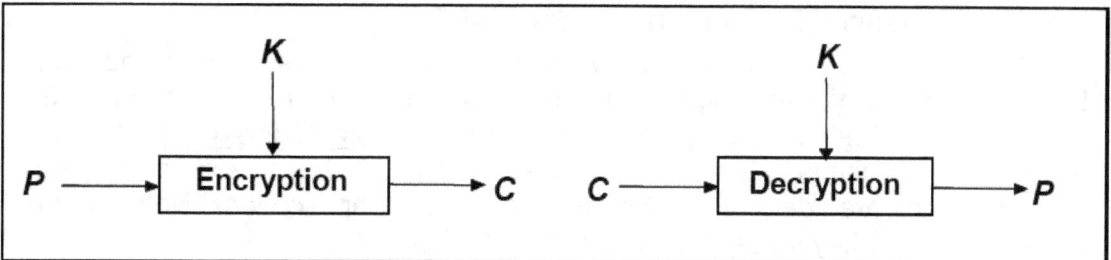

Figure 2: Encryption and Decryption

Figure 2 depicts the encryption and decryption processes. The plaintext (P) and a key (K) are used by the encryption process to produce the ciphertext (C). To decrypt, the ciphertext (C) and the same key (K) are used by the decryption process to recover the plaintext (P).

With symmetric key block cipher algorithms, the same plaintext block and key will always produce the same ciphertext block. This property does not provide acceptable security. Therefore, cryptographic modes of operation have been defined to address this problem (see Section 3.3.1.4).

3.3.1.1   Data Encryption Standard (DES)

The Data Encryption Standard (DES) became effective in July 1977. It was reaffirmed several times, but the strength of the DES algorithm is no longer sufficient to adequately protect Federal government information. Therefore, DES was withdrawn as an approved algorithm in 2005. However, DES may be continue to be used as a component function (i.e., the cryptographic engine) of the Triple Data Encryption Algorithm (TDEA).

3.3.1.2   Triple Data Encryption Algorithm (TDEA)

The Triple Data Encryption Algorithm (TDEA), also known as Triple DES, uses the DES cryptographic engine to transform data in three operations. NIST SP 800-67, *Recommendation for the Triple Data Encryption Algorithm (TDEA) Block Cipher*, specifies the TDEA block cipher algorithm. TDEA will be supported for Federal use only until 2030 (see SP 800-57).

TDEA encrypts data in blocks of 64 bits, using three keys that define a key bundle. The use of three distinctly different (i.e., mathematically independent) keys is highly recommended, since this provides the most security from TDEA; this is commonly known as three-key TDEA (3TDEA or 3TDES). Two-key TDEA (2TDEA or 2TDES), in which the first and third keys are identical, and the second key is distinctly different, is acceptable at present, but is discouraged. Other configurations of keys in the key bundle shall not be used.

3.3.1.3 Advanced Encryption Standard (AES)

The Advanced Encryption Standard (AES) was developed as a replacement for DES and is the preferred algorithm for new products. AES is specified in FIPS 197 (available at http://csrc.nist.gov/publications/fips/fips197/fips-197.pdf). AES encrypts and decrypts data in 128-bit blocks, using 128, 192 or 256 bit keys. All three key sizes are adequate for Federal Government applications. Note that the use of the higher key sizes affects algorithm performance.

3.3.1.4 Encryption Modes of Operation

With a symmetric key block cipher algorithm, the same plaintext block will always encrypt to the same ciphertext block when the same symmetric key is used. If the multiple blocks in a typical message (data stream) are encrypted separately, an adversary could easily substitute individual blocks, possibly without detection. Furthermore, certain kinds of data patterns in the plaintext, such as repeated blocks, would be apparent in the ciphertext.

Cryptographic modes of operation have been defined to address this problem by combining the basic cryptographic algorithm with variable initialization values (commonly known as initialization vectors) and feedback rules for the information derived from the cryptographic operation. The *Recommendation for Block Cipher Modes of Operation* (NIST SP 800-38A) defines modes of operation for the encryption and decryption of data using block cipher algorithms such as AES and TDEA. Another part of the recommendation (SP 800-38C) defines a mode for performing both encryption and authentication (see Section 3.3.2) in a single operation under restricted conditions. Other modes that combine the encryption and authentication operations are under consideration.

3.3.2 Message Authentication Code

Message Authentication Codes (MACs) provide an assurance of authenticity and integrity. A MAC is a cryptographic checksum on the data that is used to provide assurance that the data has not changed or been altered and that the MAC was computed by the expected party (the sender). Typically, MACs are used between two parties that share a secret key to authenticate information exchanged between those parties.

Figure 3 depicts the use of message authentication codes (MACs). A MAC (MAC_1) is computed on data (M_1) using a key (K). M_1 and MAC_1 are then saved or transmitted. At a later time, the authenticity of the retrieved or received data is checked by labeling the retrieved or received data as M_2 and computing a MAC (MAC_2) on it using the same key (K). If the retrieved or received MAC (MAC_1) is the same as the newly computed MAC (MAC_2), then it can be assumed that the retrieved or received data (M_2) is the same as the original data (M_1) (i.e., M_1 = M_2). The verifying party also knows who the sending party is because no one else knows the key.

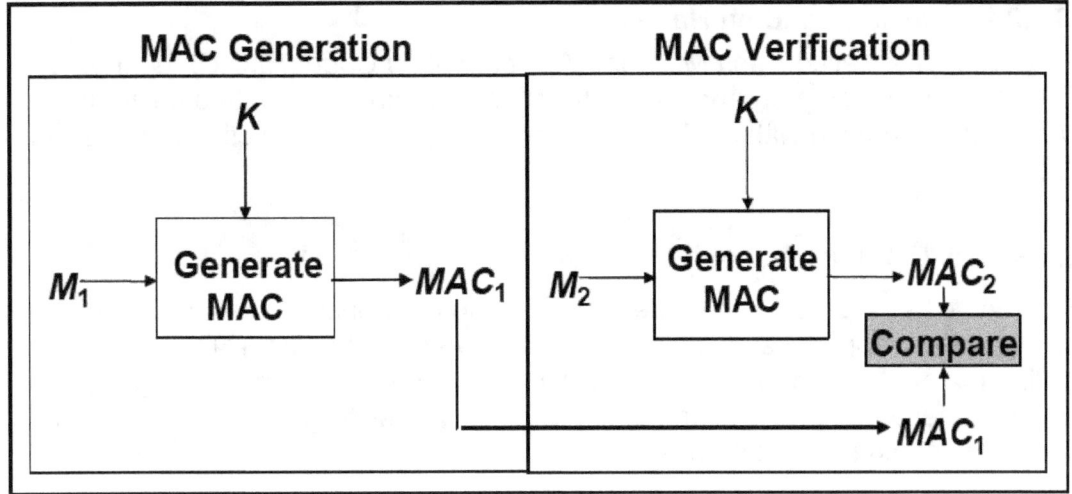

Figure 3: Message Authentication Codes (MACs)

Typically, MACs are used to detect data modifications that occur between the initial generation of the MAC and the verification of the received MAC. They do not detect errors that occur before the MAC is originally generated.

Message integrity is frequently provided using non-cryptographic techniques known as error detection codes. However, these codes can be altered by an adversary to the adversary's benefit. The use of an Approved cryptographic mechanism, such as a MAC, addresses this problem. That is, the integrity provided by a MAC is based on the assumption that it is not possible to generate a MAC without knowing the cryptographic key. An adversary without knowledge of the key will be unable to modify data and then generate an authentic MAC on the modified data. It is therefore crucial that MAC keys be kept secret.

Two types of algorithms for computing a MAC have been approved for Federal government use: MAC algorithms that are based on block cipher algorithms, and MAC algorithms that are based on hash functions.

3.3.2.1 MAC Based on a Block Cipher Algorithm

NIST SP 800-38B, *Recommendation for Block Cipher Modes of Operation: the CMAC Authentication Mode*, defines a mode to compute a MAC using approved block cipher algorithms, such as AES and TDEA. If the same block cipher algorithm is used for both encryption and MAC computation (i.e., using a mode from SP 800-38A for encryption, and a mode from SP 800-38B for MAC computation), then different keys shall be used for each operation. SP 800-38C, however, defines a mode for performing both encryption and MAC computation in a single operation and using a single key under restricted conditions.

3.3.2.2 MACs Based on Hash Functions

FIPS 198, *The Keyed Hash Message Authentication Code (HMAC)*, defines a MAC that uses a cryptographic hash function in combination with a secret key. HMAC shall be used with an Approved cryptographic hash function (see Section 3.2).

3.3.3 Key Establishment

Symmetric key algorithms may be used to wrap (i.e., encrypt) keying material using a key-wrapping key (also known as a key encrypting key). The wrapped keying material can then be stored or transmitted securely. Unwrapping the keying material requires the use of the same key-wrapping key that was used during the original wrapping process.

Key wrapping differs from simple encryption in that the wrapping process includes an integrity feature. During the unwrapping process, this integrity feature detects accidental or intentional modifications to the wrapped keying material.

There is currently no FIPS or NIST-recommendation that specifies the key wrapping algorithm, but a specification for an algorithm using AES is available at http://csrc.nist.gov/CryptoToolkit/tkkeymgmt.html. The AES key wrapping algorithm is anticipated to be specified in a future part of SP 800-38 as part D, and may include an adaption of its use for TDEA in addition to AES.

A future publication is also under development that will specify additional guidance for key establishment using symmetric key techniques.

3.4 Asymmetric Key Algorithms

Asymmetric algorithms (often called public key algorithms) use two keys: a public key and a private key, which are mathematically related to each other. The public key may be made public; the private key must remain secret if the data is to retain its cryptographic protection. Even though there is a relationship between the two keys, the private key cannot be determined from the public key. Which key to be used to apply versus remove or check the protection depends on the service to be provided. For example, a digital signature is computed using a private key, and the signature is verified using the public key; for those algorithms also capable of encryption[8], the encryption is performed using the public key, and the decryption is performed using the private key.

Asymmetric algorithms are used primarily as data integrity, authentication, and non-repudiation mechanisms (i.e., digital signatures), and for key establishment.

---

[8] Not all public key algorithms are capable of multiple functions, e.g., generating digital signatures and encryption.

Some asymmetric algorithms use domain parameters, which are additional values necessary for the operation of the cryptographic algorithm. These values are mathematically related to each other. Domain parameters are usually public and are used by a community of users for a substantial period of time.

The secure use of asymmetric algorithms requires that users obtain certain assurances:

- Assurance of domain parameter validity provides confidence that the domain parameters are mathematically correct,

- Assurance of public key validity provides confidence that the public key appears to be a suitable key, and

- Assurance of private key possession provides confidence that the party that is supposedly the owner of the private key really has the key.

Some asymmetric algorithms may be used for multiple purposes (e.g., for both digital signatures and key establishment). Keys used for one purpose shall not be used for other purposes.

3.4.1 Digital Signatures and the Digital Signature Standard (DSS)

A digital signature is an electronic analogue of a written signature that can be used in proving to the recipient or a third party that the message was signed by the originator (a property known as non-repudiation). Digital signatures may also be generated for stored data and programs so that the integrity of the data and programs may be verified at a later time.

Digital signatures authenticate the integrity of the signed data and the identity of the signatory. A digital signature is represented in a computer as a string of bits and is computed using a digital signature algorithm that provides the capability to generate and verify signatures. Signature generation uses a private key to generate a digital signature. Signature verification uses the public key that corresponds to, but is not the same as, the private key to verify the signature. Each signatory possesses a private and public key pair. Signature generation can be performed only by the possessor of the signatory's private key. However, anyone can verify the signature by employing the signatory's public key. The security of a digital signature system is dependent on maintaining the secrecy of a signatory's private key. Therefore, users must guard against the unauthorized acquisition of their private keys.

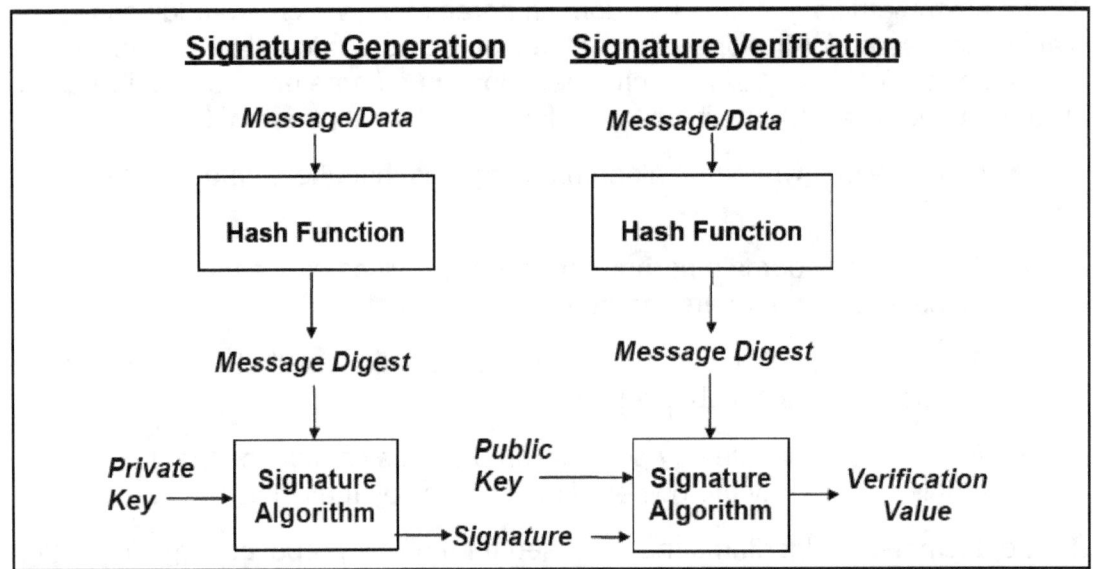

Figure 4: Digital Signatures

Figure 4 depicts the digital signature process. A hash function (see Section 3.2) is used in the signature generation process to obtain a condensed version of data to be signed, called a message digest or hash value. The message digest is then input to the digital signature algorithm to generate the digital signature. The digital signature is sent to the intended verifier along with the signed data (often called the message). The verifier of the message and signature verifies the signature by using the signatory's public key. The same hash function and digital signature algorithm must also be used in the verification process. Similar procedures may be used to generate and verify signatures for stored as well as transmitted data.

Digital signatures offer protection that is not available by using alternative signature techniques. One such alternative is a digitized signature. A digitized signature is generated by converting a visual form of a handwritten signature to an electronic image. Although a digitized signature resembles its handwritten counterpart, it does not provide the same protection as a digital signature. Digitized signatures can be forged and can be duplicated and appended to other electronic data; digitized signatures cannot be used to determine if information has been altered after it is signed. Digital signatures, however, are computed on each message using a private key known only by the signatory. Each different message signed by the signatory will have a different digital signature. Even small changes to the message will result in a different signature. If an adversary does not know the private key, a valid signature cannot be generated.

FIPS 186-3, *Digital Signature Standard (DSS)*, specifies methods for generating and verifying digital signatures using asymmetric (public key) cryptography. DSS includes three digital signature algorithms: the Digital Signature Algorithm (DSA), the Elliptic Curve Digital Signature Algorithm (ECDSA) and RSA. The DSS is used in conjunction with FIPS 180-2, *Secure Hash Standard.*

FIPS 186-3 specifies methods for the generation of domain parameters and private/public key pairs, the selection of key sizes and hash functions, and the generation and verification of digital signatures. The standard also provides methods for obtaining assurances of domain parameter validity, public key validity, and possession of the private key. A method for generating random numbers used to generate secret values (e.g., keys) is also provided.

3.4.2 Key Establishment

Two types of asymmetric key (i.e., public key) establishment are defined: key transport and key agreement. Approved key establishment schemes are specified in NIST SP 800-56, Recommendation on Key Establishment Schemes.

Key transport is the distribution of a key (and other keying material) from one party to another party. The transported key is created by the sending party. The keying material is encrypted by the sending party and decrypted by the receiving party. The sending party encrypts the keying material using the receiving party's public key; the receiving party decrypts the received keying material using the associated private key.

Key agreement is the participation by both parties (i.e., the sending and receiving parties) in the creation of shared keying material. Each party has either one or two key pairs[9], and the public keys are made known to the other party. The key pairs are used to compute a shared secret[10] value, which is then used with other information to derive keying material using a key derivation function. Typically, a hash function (see Section 3.1) is used during the key derivation process.

NIST SP 800-56 specifies selected key establishment schemes: Diffie-Hellman and MQV schemes using two different types of mathematics, finite field and elliptic curve. In the future, RSA schemes will also be included.

NIST SP 800-56 includes discussions on:

- Domain parameter generation and assurance of domain parameter validity,
- Public and private key pair generation and assurance of public key validity and private key possession,
- Several key establishment schemes and associated functions,
- Methods for providing key transport, and

---

[9] A key pair consists of a private key and its associated public key.

[10] The shared secret is never transmitted from one party to the other.

- Techniques for providing key confirmation to obtain assurance that the participating parties share the same keying material.

3.5 Random Number Generation

Random number generators (RNGs) are required for the generation of keying material (e.g., keys and initialization vectors (IVs)). There are two classes of RNGs: deterministic and non-deterministic. Deterministic Random Bit Generators (DRBGs), sometimes called deterministic random number generators or pseudorandom number generators, use cryptographic algorithms to generate random numbers. Non-Deterministic Random Bit Generators (NRBGs), sometimes called true RNGs, produce output that is dependent on some unpredictable physical source that is outside human control, for example, radioactive decay or a true noise hardware randomizer.

FIPS 186-3 defines a DRBG that may be used to generate random numbers for cryptographic applications (e.g., key or IV generation). The DRBG is initialized with a secret starting value, called an RNG seed, and uses a hash function.

Further guidance on the design and use of random number generators is under development.

3.6 Key Management

The proper management of cryptographic keys is essential to the effective use of cryptography for security. Keys are analogous to the combination of a safe. If a safe combination becomes known to an adversary, the strongest safe provides no security against penetration. Similarly, poor key management may easily compromise strong algorithms. Ultimately, the security of information protected by cryptography directly depends on the strength of the keys, the effectiveness of mechanisms and protocols associated with keys, and the protection afforded to the keys. All keys need to be protected against modification, and secret and private keys need to be protected against unauthorized disclosure. Key management provides the foundation for the secure generation, storage, distribution, and destruction of keys.

NIST Special Publication 800-57 (SP 800-57), *Recommendation for Key Management*, provides guidance on the management of cryptographic keys: their generation, use, and eventual destruction. Related topics, such as algorithm selection and appropriate key size, cryptographic policy, and cryptographic module selection, are also included in this recommendation. SP 800-57 consists of three parts:

- Part 1, *General Guidance*, contains basic key management guidance, including:
    - the protection required for keying material,
    - the key life cycle responsibilities,

- o key backup, archiving and recovery,
- o changing keys,
- o cryptoperiods (i.e., the appropriate lengths of time that keys are to be used),
- o accountability and audit,
- o contingency planning and
- o key compromise recovery.
- Part 2, *Best Practices for Key Management Organizations*, contains:
  - o A generic key management infrastructure,
  - o Guidance for the development of organizational key management policy statements and key management practices statements,
  - o An identification of key management information that needs to be incorporated into security plans for general support systems and major applications that employ cryptography, and
  - o An identification of key management information that needs to be documented for all Federal applications of cryptography.
- Part 3, *Application-Specific Key Management Guidance*, addresses the key management issues associated with currently available cryptographic mechanisms, such as the Public Key infrastructure (PKI), the Transport Layer Security protocol (TLS), and Secure/Multipart Internet Mail Extensions (S/MIME). Specific guidance is provided regarding:
  - o The recommended and/or allowable algorithm suites and key sizes,
  - o Recommendations for the use of the mechanism in its current form for the protections of Federal government information, and/or
  - o Security considerations that may affect the security effectiveness of key management processes and the cryptographic mechanisms that use keys that are generated and managed by those key management processes.

New key management techniques and mechanisms are constantly being developed, and existing key management mechanisms and techniques are constantly being refined. While the security guidance information contained in Part 3 will be updated as mechanisms and techniques evolve, new products and technical specifications can always be expected that are not reflected in the current version of the guideline. Therefore, the context provided may include status information, such as version numbers or implementation status.

Additional key management guidance is provided in FIPS 140-2, *Security Requirements for Cryptographic Modules*, which provides minimum security requirements for cryptographic modules that embody or support key management in Federal information systems.

Federal agencies have a variety of information that they have determined to require cryptographic protection; the sensitivity of the information and the periods of time that the protection is required also vary. To this end, NIST has established five security levels (i.e., security strengths) for the protection of information: 80, 112, 128, 192 and 256. These security levels have been assigned to the Approved cryptographic algorithms and key sizes, and dates have been projected during which the use of these algorithms and key sizes is anticipated to be secure. For further information, see SP 800-57.

Agencies need to determine the length of time that cryptographic protection is required before selecting an algorithm and key size with the appropriate cryptographic strength.

3.7 Public Key Infrastructure (PKI)[11]

A PKI is a security infrastructure that creates and manages public key certificates to facilitate the use of public key (i.e., asymmetric key) cryptography. To achieve this goal, a PKI needs to perform two basic tasks:

1. Generate and distribute public key certificates to bind public keys to other information *after* validating the accuracy of the binding; and

2. Maintain and distribute certificate status information for unexpired certificates.

Some aspects of these tasks are relevant to the trustworthiness of the PKI. Other aspects affect the availability and performance of the PKI. The core tasks of the PKI are binding public keys to accurate information in a digitally signed certificate, and maintaining accurate certificate status information. If the components that implement these core tasks are implemented poorly, the PKI itself may be compromised. The distribution of certificates and status information affects the utility and performance of a PKI. If the components that handle distribution are compromised, denial of service may result, but the trustworthiness of the PKI is unaffected.

The use of certificates ensures the availability of the public keys. However, a private keys is maintained under the exclusive control of the owner of that private key (i.e., the user that is authorized to use the private key). A user can only operate within a PKI if his private key is readily available.

---

[11] Information in this section was extracted from:
http://csrc.nist.gov/pki/documents/CIMC_PP_20011031.pdf

If a private key that is used to generate digital signatures is lost, the owner can no longer generate digital signatures. Policy may permit users to maintain backup copies of the private key for their own convenience, but continuity of operations may be achieved by simply generating new key pairs and certificates.

If a private key used for key management is lost (e.g., a key used for key transport or key agreement), then access to the data protected by that key may no longer be possible. For example, if the key is used to transport a decryption key for encrypted data, and the key management key is lost, then the encrypted data cannot be decrypted. To ensure that access to critical data is not lost, PKIs often backup the private key management key for possible recovery. While key recovery is orthogonal to the main goal of a PKI (i.e., the distribution of public keys), the trustworthiness of a PKI may depend greatly upon the security of this backup/recovery function. Securely implemented key recovery services will enhance the utility and dependability of PKI-based applications, but an insecure implementation will compromise the confidentiality of any PKI-dependent application.

Even where users maintain control of the private keys, the PKI may provide centralized storage to support mobile, or roaming, users. When roaming users wish to perform cryptographic operations, they obtain their credentials (e.g., private keys and their corresponding certificates) and perform the cryptographic operations on whatever workstation they currently are using. As above, roaming users generally have exclusive control of digital signature keys, but the PKI may maintain copies of the private keys used for key establishment.

A monolithic PKI component could be designed to satisfy all of these requirements, but this is not common practice. For scalability, PKIs are usually implemented with a set of complementary components, each focused on specific aspects of the PKI process. The PKI tasks are often assigned to the following logical components:

- *Certification authorities* (CAs) to generate certificates and certificate status information;
- *Registration authorities* (RAs) to verify the information in the public key certificates and determine certificate status;
- *Repositories* to distribute certificates and certificate revocation lists (CRLs);
- *Online Certificate Status Protocol* (OCSP) *servers* to distribute certificate status information in the form of OCSP responses;
- *Key recovery servers* to backup private key material; and
- *Roaming credential servers* to distribute private key material and the corresponding certificates.

A particular PKI implementation must include the functionality of CAs and RAs, but the requirements may be assigned to any number of components. The features provided by repositories, OCSP servers, key recovery servers, and roaming credential servers are optional in a PKI implementation.

3.7.1 Security Requirements for PKI Components

The Certificate Issuing and Management Components (CIMC) Family of Protection Profiles defines requirements for components that issue, revoke, and manage public key certificates, such as X.509 public key certificates. A CIMC always includes a Certification Authority (CA) and may include Registration Authorities (RAs) and other subcomponents.

A CIMC consists of the hardware, software, and firmware that are responsible for issuing, revoking, and managing public key certificates. A CIMC does not include environmental controls (e.g., controlled access facility, temperature), policies and procedures, personnel controls (e.g., background checks and security clearances), and other administrative controls.

The _Certificate Issuing and Management Component (CIMC) Family of Protection Profile_s specifies the functional and assurance security requirements for a CIMC. The intent of this family of Protection Profiles is to ensure specification of the complete set of requirements for a CIMC and not the specification of a subset of requirements implemented in a specific CIMC subcomponent. It includes all the technical features of a CIMC, regardless of which CIMC subcomponent performs the function. The document does not differentiate between functions that are typically performed by a CA and functions that are typically performed by an RA.

3.7.2 PKI Architectures

A PKI often includes many CAs linked by trust paths. The CAs may be linked in several ways. They may be arranged hierarchically under a "root CA" that issues certificates to subordinate CAs. The CAs can also be arranged independently in a mesh[12]. Recipients of a signed message with no relationship with the CA that issued the certificate for the message sender can still validate the sender's certificate by finding a path between their CA and the one that issued the sender's certificate. Figures 5 and 6 illustrate the two basic PKI architectures.

In hierarchical models, trust is delegated by a CA when it certifies a subordinate CA. Trust delegation starts at a root CA that is trusted by every node in the infrastructure. In mesh models, trust is established between any two CAs in peer relationships (cross-certification), thus allowing the possibility of multiple trust paths between any two CAs. Note that hierarchical and mesh PKIs are not mutually exclusive, and may be combined into more complex PKIs. For example, cross certifying the root CA from a hierarchy with any CA in a mesh PKI creates a new PKI that exhibits aspects of both architectures.

---

[12] A mesh PKI model is sometimes referred to as a _network_ PKI model.

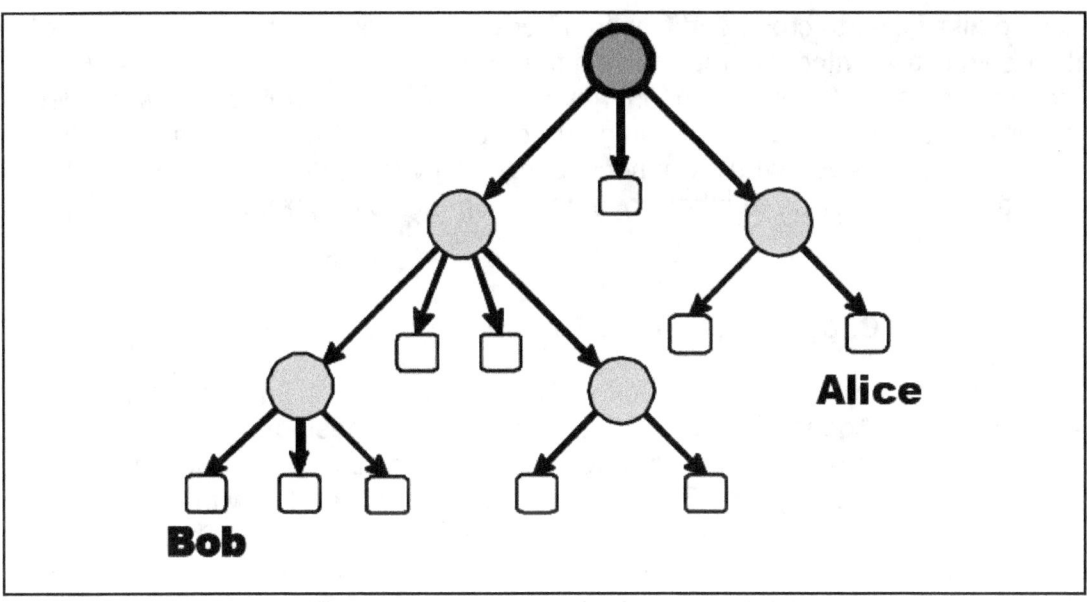

Figure 5. Hierarchical Architecture

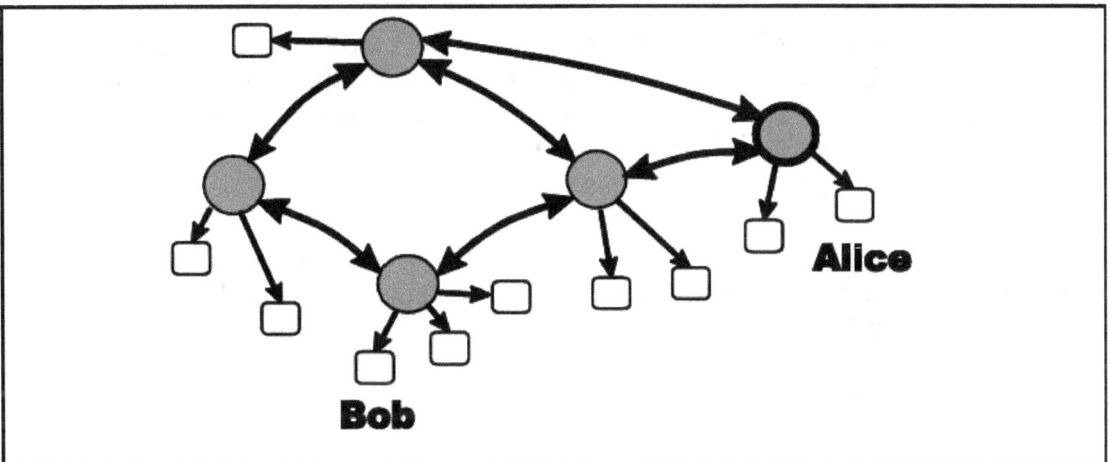

Figure 6. Mesh Architecture

3.7.3 Security Policies of Other CAs and the Network

It is important to consider the integrity and security of the PKI components. The confidence that can be placed in the binding between a public key and its owner depends, in large part, on the confidence that can be placed on the system that issued the certificate that binds them. The rules expressed by certificate policies are reflected in certification practice statements (CPSs) that detail the operational rules and system features of CAs and other PKI components. By examining a CA's CPS, users can determine whether to obtain certificates from it, based on their security requirements. Other CAs can also use the CPS to determine if they want to cross-certify with that CA. The essential issue with cross-certificates is how to allow CAs to cross-certify with other CAs to meet the particular needs of their own users, without compromising the security of users of other CAs. For example, a particular agency might have a close working relationship with a local government office, a particular contractor or law firm that has its own CA. That relationship, however, would not necessarily justify the extension of trust by that local government office to other government agencies or commercial organizations.

3.7.4 Federal Bridge Certification Authority

The Federal Bridge Certification Authority (FBCA) supports interoperability among Federal Agency PKI domains in a peer-to-peer fashion. The FBCA will issue a certificate only to those Agency CAs specified by the owning Agency (called "Principal CAs"). The FBCA, or a CA that interoperates with the FBCA, may also issue certificates to individuals who operate the FBCA. The FBCA certificates issued to Agency Principal CAs act as a conduit of trust. The FBCA does not add to and should not subtract from trust relationships existing between the transacting parties. The Federal PKI Policy Authority (FPKIPA) is the governing body over the FBCA.

The X.509 Certificate Policy for the FBCA defines five certificate policies for use by the FBCA to facilitate Agency CA interoperability with the FBCA and with other Agency PKI domains. The five policies represent four different assurance levels (Rudimentary, Basic, Medium, and High) for public key digital certificates, plus one assurance level used strictly for testing purposes (Test). The word "assurance" used in this CP means how well a Relying Party[13] can be certain of the identity binding between the public key and the individual whose subject name is cited in the certificate. In addition, it also reflects how well the Relying Party can be certain that the individual whose subject name is cited in the certificate is controlling the use of the private key that corresponds to the public key in the certificate.

---

[13] In general, a Relying Party is a recipient that relies on the validity of a certificate and digital signature.

CHAPTER 4

GENERAL IMPLEMENTATION ISSUES

There are many issues that are applicable to the implementation of security methods/products. These are extensively discussed in other documents such as:

- *Guide for Developing Security Plans for Information Technology Systems* (NIST SP-800-18),

- *Guide for the Security Certification and Accreditation of Information Technology Systems* (NIST SP 800-37),

- *Recommendation for Key Management* (NIST SP 800-57),

- OMB Circular A-130, *Security of Federal Automated Information Resources*, Appendix III,

- *Recommended Security Controls for Federal Information Systems* (NIST SP-800-53), and

- *Personal Identity Verification for Federal Employees and Contractors* (FIPS 201).

4.1 Hardware vs. Software Solutions

Agencies need to evaluate trade-offs among security, cost, simplicity, efficiency, and ease of implementation. Cryptography can be implemented in hardware, software and/or firmware - each has its related costs and benefits.

Historically, software has been less expensive and slower than hardware, although for large applications, hardware may be less expensive. In addition, software is easier to modify or bypass[14] than equivalent hardware products. The protection of key variables upon which cryptographic security depends is also more difficult to achieve in software-based implementations. The advantages of software solutions are in flexibility and portability, ease of use, and ease of upgrade.

In many cases, cryptography is implemented in a hardware device but is controlled by software and, therefore, a hybrid solution is provided. Again, the user must evaluate the solutions against requirements to determine the best solution.

---

[14] Note that this can be a security concern or an advantage (e.g., when there are problems with cryptographic functionality).

4.2 Asymmetric vs. Symmetric Cryptography

Symmetric, or secret key, cryptography employs a single key to both apply cryptographic protection (e.g., encrypt) and to remove or check the protection (e.g., decrypt). This key must be kept secret if the underlying cryptographic process is to be effective. Symmetric cryptography is discussed in Section 3.3.

Asymmetric, or public key, cryptography employs interdependent pairs of keys: a key that may be made public and a key that must remain private. Asymmetric cryptography is discussed in Section 3.4.

When keying material needs to be provided to other entities for cryptographic relationships, symmetric and asymmetric cryptography differ as follows:

- *Symmetric cryptography:* A unique key needs to be generated for each relationship[15] and for each purpose (e.g., encryption, authentication and key wrapping). For example, if there are four entities (A, B, C, and D) using encryption, there are six possible relationships (A-B, A-C, A-D, B-C, B-D, C-D). If a key is to be provided for encryption for each relationship, six keys are required. If there are, instead, 1000 entities there are 499,500 possible relationships, and a unique key would be required for each relationship. The method for transferring the key from the sending party to each recipient must provide for both confidentiality and data integrity protection for the key.

- *Asymmetric cryptography*: A private/public key pair needs to be generated by each party that needs a private key to sign data, a separate key pair for each type of key agreement process, and a separate key pair to receive transported keys. For example, four entities need four key pairs for digital signatures, and 1000 entities need 1000 key pairs. A unique key does not need to be generated for each relationship.

   The private key is retained by the entity who "owns" the key pair and must be kept secret. The public key is distributed to the other entities and requires integrity protection (e.g., using a public key certificate prior to providing the public key to other entities). This integrity protection may be the same for all the relationships (e.g., a single public key certificate can be provided to all interested parties).

Therefore, for networks that have large numbers of pair-wise relationships, the number of symmetric keys that will require confidentiality protection is significantly larger than the number of public/private key pairs.

The primary advantage of symmetric cryptography is speed. There are approved symmetric key algorithms that are significantly faster than any currently available asymmetric key algorithm. In addition, advances in factoring efficiency, other

---

[15] A relationship may be one-to-one or one-to-many (e.g., broadcast).

cryptographic methods, and computational efficiency have tended to reduce the protection provided by public key cryptography more rapidly than that provided by symmetric key cryptography.

Asymmetric and symmetric cryptography can be used together to obtain the key management advantages of public-key systems and the encryption speed advantages of symmetric systems. For example, an asymmetric system can be used to transport symmetric keys that are used to encrypt files or messages.

In some situations, asymmetric cryptography is not necessary, and symmetric cryptography alone is sufficient. This includes environments where secure symmetric key establishment can take place (see Section 3.3.3), environments where a single authority knows and manages all the keys, and a single-user environment. In general, asymmetric cryptography is best suited for an open multi-user environment.

4.3 Key Management

The proper management of cryptographic keys is essential to the effective use of cryptography for security. Ultimately, the security of information protected by cryptography directly depends on the protection afforded the keys. All keys need to be protected against modification, and secret and private keys need to be protected against unauthorized disclosure.

NIST and other standards organizations have produced guidelines for effective key management. The *Recommendation for Key Management* (NIST SP 800-57; discussed in Section 3.6) is a three part general guide to key management.

Listed below are some general recommendations for effective key management.

Make sure that users are aware of their liabilities and responsibilities, and that they understand the importance of keeping their keys secure.

The security of cryptographic keys is the foundation of a secure system; therefore, users must maintain control of their keys! Users must be provided with a list of responsibilities and liabilities, and each user should sign a statement acknowledging these concerns before receiving a key (if it is a long-term, user-controlled key). If different user roles (e.g., security officer, regular user) are implemented in a system, users shall be made aware of their unique responsibilities, especially regarding the significance of a key compromise or loss.

Prepare for a possible compromise

It is imperative to have a plan for handling the compromise or suspected compromise of central/root keys or key components at a central site; this should be established before the system goes operational. A contingency plan should address what actions will be taken with compromised system software and hardware, central/root keys, user keys, previously generated signatures, encrypted data, etc.

If someone's private or secret key is lost or compromised, other users shall be made aware of this, so that they will no longer initiate the protection of messages using a compromised key nor accept messages protected with a compromised key. Users must be able to store their secret and private keys securely, so that no intruder can access them, yet the keys must be readily accessible for legitimate use.

Use validated algorithms and modules.

Cryptographic algorithms and the cryptographic modules in which they reside shall be validated in accordance with FIPS 140-2.

Software at a central key management site should be electronically signed and periodically verified by the user's system to check the integrity of the code. This provides a means of detecting the unauthorized modification of system software. Within a cryptographic module, the generation and verification of a cryptographic checksum is required by FIPS 140-2.

A system implemented for a Federal government agency should have centrally stored keys and system components controlled by Federal employees.

The proper control of central/root keys and key management components is critical to the security of a system. When a Federal system is developed by a contractor, Federal employees should control the keying material and configure the key management components. Once the system becomes operational, unlimited access to central data, code, and cryptographic modules should not be given to non-Federal employees.

Secure Key Management

Key management provides the foundation for the secure generation, storage, distribution, and use of keys.

Proper key management is essential at all phases of the keys' life. Guidance for the management of cryptographic keying material is provided in NIST SP 800-57. Additional guidance includes the following:

> *Provide a cryptographic integrity code (e.g., a digital signature or MAC) for all centrally stored data and encrypted sensitive data, such as secret keys that are used to provide confidentiality.*

> All centrally stored data that is related to user keys should be signed or MACed for integrity, and encrypted if confidentiality is required (all user secret keys and CA private keys should be encrypted). Individual key records in a database - as well as the entire database - should be signed or MACed and encrypted. To enable tamper detection, each individual key record should be

signed or MACed so that its integrity can be checked before allowing that key to be used in a cryptographic function.

Provide back-up copies of keys.

Backup copies should be made of central/root keys, since the compromise or loss of those components could prevent access to keys in the central database, and possibly deny system users the ability to decrypt data or perform signature verifications.

Provide key recovery capabilities.

IT systems shall protect the confidentiality of information. There must be safeguards to ensure that sensitive records are neither irretrievably lost by the rightful owners nor accessed by unauthorized individuals. Key recovery capabilities provide these functions. All key components used for encryption should be available to an organization, regardless of whether the associated user is currently working in the organization. Employees leave organizations voluntarily and some are removed. In either situation, the organization may need to access the keys to recover encrypted data. Key recovery capabilities allow organizations to restore key components.

Archive user keys for a sufficiently long cryptoperiod.

A cryptoperiod is the time during which a key can be used to protect information; it may extend well beyond the lifetime of a key that is used to apply cryptographic protection (where the lifetime is the time during which a key can be used to generate a signature and/or perform encryption). Keys may be archived for a lengthy period (on the order of decades), so that they can be used to verify signatures and decrypt ciphertext.

Determine reasonable lifetimes for keys associated with different types of users.

Users with different roles should have keys with lifetimes that take into account the different roles and responsibilities, the applications for which the keys are used, and the security services that are provided by the keys (user/data authentication, confidentiality, data integrity, etc.). Reissuing keys should not be done so often that it becomes excessively burdensome; however, it should be performed often enough to minimize the loss caused by a possible key compromise.

Handle the deactivation/revocation of keys so that data signed prior to a compromise date (or date of loss) can be verified.

When a signing key is designated as "lost" or "compromised," signatures generated prior to the specified date may still need to be verified in the future. Therefore, a signature verification capability may need to be maintained for lost/compromised keys. Otherwise, all data previously signed with a lost/compromised key would have to be re-signed.

CHAPTER 5
ASSESSMENTS

Cryptographic controls are provided using cryptographic modules, which may include capabilities such as signature generation and verification, encryption and decryption, key generation, and key establishment.

An undetected error in a cryptographic module's design could affect every user in the system for which it is supposed to provide protection. For example, the verification of a chain of public key certificates might not function correctly. Verifying a chain of public key certificates helps a signature verifier determine if a signature was generated with a particular key. If the function is implemented incorrectly in a cryptographic module, the potential for the dissemination of weak cryptography could be introduced into the system, possibly allowing for signature forgery or the verification of invalid signatures. Therefore, it is important to have cryptographic modules tested before distributing them throughout a system.

Figure 7 illustrates a general security testing model, including the testing of cryptographic modules, and the various types of testing that are required. This model, and the applicable testing organizations, is described in this chapter.

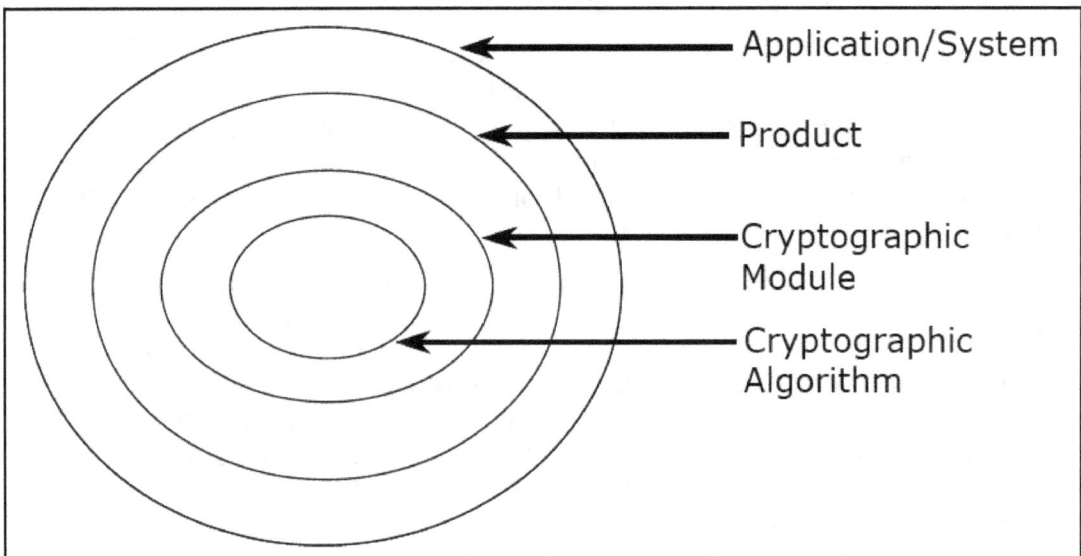

Figure 7: Security Testing Model

Table 2 illustrates the range of cryptographic tests, from individual algorithms to complete systems.

Table 2: Uses of Tests

Test	Example	Specification
Application/System	Air Traffic Control	Certification and Accreditation (SP 800-37)
Product	Security Module	Common Criteria Evaluation and Validation Scheme (CCEVS)
Security Module	Crypto Module	FIPS 140-2[16]
Algorithm	AES	FIPS 197

Cryptographic algorithms and cryptographic modules must be tested prior to integration into an existing or new system. The cryptographic algorithms and modules are tested by the developer and then submitted to the Cryptographic Module Validation Program (CMVP) for testing against FIPS 140-2, *Security Requirements for Cryptographic Modules* and the applicable cryptographic algorithm standards.

For all Federal agencies, the use of cryptographic products that conform to FIPS 140-2 is mandatory for the protection of sensitive unclassified information when the agency determines that cryptographic protection is required[17]. Agencies are required to use FIPS 140-2 in designing, acquiring, and implementing cryptographic-based security systems within computer and telecommunications systems (including voice systems).

5.1 Cryptographic Module Validation Program (CMVP)

NIST and the Communications Security Establishment (CSE) of the government of Canada established the CMVP. The goal of the CMVP is to provide Federal agencies with a security metric to use in procuring equipment containing cryptographic modules. The results of the independent testing by accredited laboratories provide this metric. Cryptographic module validation testing is performed using the Derived Test Requirements (DTRs) for FIPS 140-2. The DTRs list all of the vendor and tester requirements for validating a cryptographic module and are the basis of testing done by the Cryptographic Module Testing (CMT) accredited laboratories.

---

[16] Note that FIPS 140-2 is being updated.

[17] National security-related information is excluded from this requirement.

5.1.1 Background

A cryptographic module is a set of hardware, firmware or software, or some combination thereof, that implements cryptographic logic or processes. Examples include standalone devices such as link encryptors; add-on encryption boards embedded in computer systems; and software applications running on microprocessors, such as digital signature applications. If the cryptographic logic is implemented in software, then the processor, which executes the software, is also part of the cryptographic module.

There are many advantages to using validated modules:

- Assurance that modules incorporate necessary features,

- Protection of technical assets and staff time of government personnel by assuring that purchased products comply with a standard and have been tested,

- Providing users with a set of available and relevant security features, and

- Increased flexibility to choose security requirements that meet application-specific and environment-specific requirements.

Figure 8 illustrates the CMV process. The process begins with the submission of the cryptographic module to one of the accredited laboratories. During the testing process, there are typically many interactions between the laboratory and the vendor and between the laboratory and NIST/CSE. NIST/CSE respond to questions about a specific validation and issue general implementation guidance that is applicable to all validations. The implementation guidance is not static, and is augmented as needed to respond to questions. The laboratory then writes the test report and submits it to NIST/CSE for validation. NIST/CSE review the test report and request clarification from the laboratory, as required. Finally, NIST/CSE issue the validation certificate and update the CMVP web site[18]. Note that the web site contains a list of validated modules (see 5.1.4).

In general, NIST/CSE responsibilities include:

- Reviewing reports and issuing validation certificates,

---

[18] FIPS 140-2, DTRs, implementation guidance, and pre-validated and validated modules lists are located at the web site: www.nist.gov/cmvp.

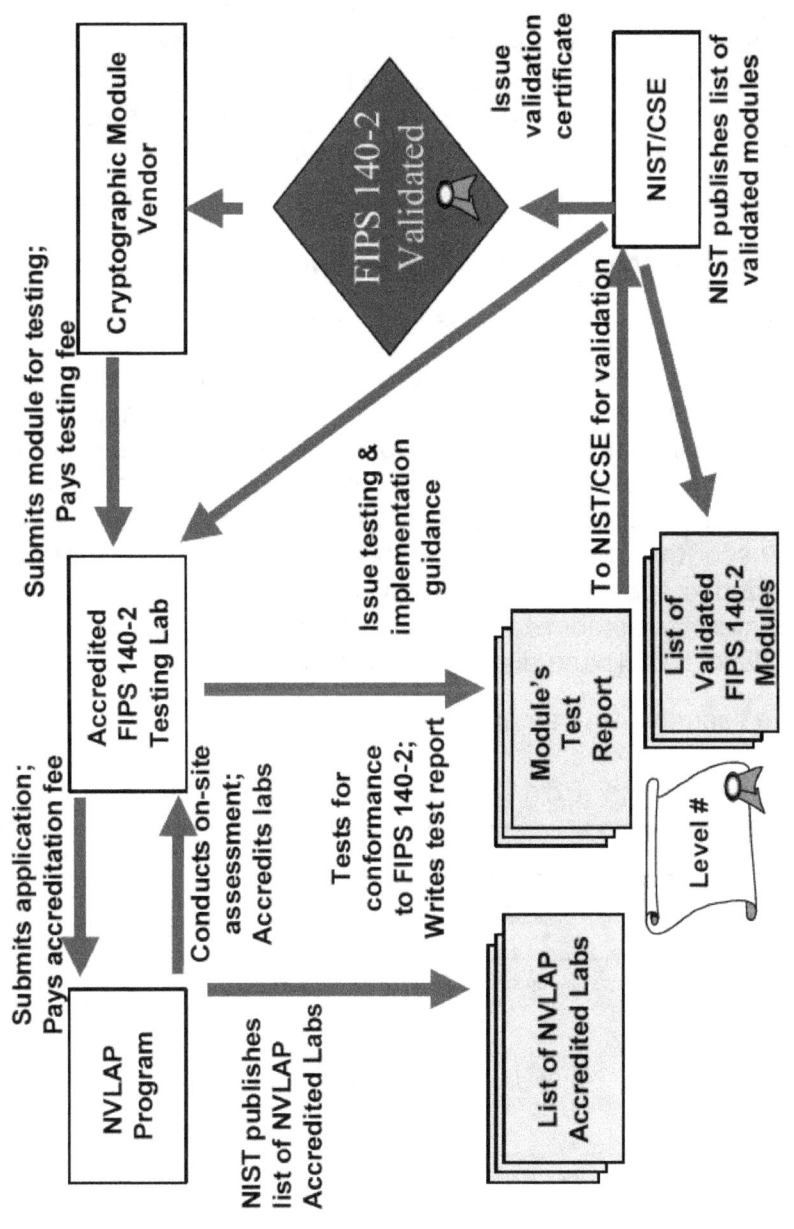

Figure 8: CMVP Processes

- Issuing CMVP policies,
- Issuing guidance and clarifications of FIPS 140-2 and other cryptography standards (to labs, vendors, government organizations, and others), and
- Assisting the National Voluntary Laboratory Accreditation Program (NVLAP) in laboratory assessments.

5.1.2 FIPS 140-2 Requirements

The security requirements in FIPS 140-2 cover 11 areas related to the design and implementation of a cryptographic module. Within most areas, a cryptographic module receives a security level rating of 1 to 4, from lowest to highest, depending on what requirements are met. For other areas that do not provide for different levels of security, a cryptographic module receives a rating that reflects the fulfillment of all of the requirements for that area.

An overall rating is issued for the cryptographic module, that indicates the:

1. Minimum of the independent ratings received in the areas with levels, and
2. Fulfillment of all the requirements in the other areas.

On a vendor's validation certificate, individual ratings are listed as well as the overall rating. It is important for vendors and users of cryptographic modules to realize that the overall rating of a cryptographic module is not necessarily the most important rating. The rating of an individual area may be more important than the overall rating, depending on the environment in which the cryptographic module will be used (this includes understanding what risks the cryptographic module is intended to address). Modules may meet different levels in different security requirement areas; for example, a module may implement identity-based authentication (level 3 or 4) and display tamper evidence (level 2).

Table 3 lists the security requirements from FIPS 140-2.

Table 3: FIPS 140-2 Security Requirements

	Security Level 1	Security Level 2	Security Level 3	Security Level 4
Cryptographic Module Specification	Specification of cryptographic module, cryptographic boundary, Approved algorithms, and Approved modes of operation. Description of cryptographic module, including all hardware, software, and firmware components. Statement of module security policy.			
Cryptographic Module Ports and Interfaces	Required and optional interfaces. Specification of all interfaces and of all input and output data paths.		Data ports for unprotected critical security parameters logically or physically separated from other data ports.	
Roles, Services, and Authentication	Logical separation of required and optional roles and services.	Role-based or iden ity-based operator authentication.	Iden ity-based operator authentication.	
Finite State Model	Specification of finite state model. Required states and optional states. State transi ion diagram and specification of state transitions.			

	Security Level 1	Security Level 2	Security Level 3	Security Level 4
Physical Security	Production grade equipment.	Locks or tamper evidence.	Tamper detection and response for covers and doors.	Tamper detection and response envelope. EFP or EFT.
Operational Environment	Single operator. Executable code. Approved integrity technique.	Referenced PPs evaluated at EAL2 with specified discretionary access control mechanisms and auditing.	Referenced PPs plus trusted path evaluated at EAL3 plus security policy modeling.	Referenced PPs plus trusted path evaluated at EAL4.
Cryptographic Key Management	Key management mechanisms: random number and key generation, key establishment, key distribution, key entry/output, key storage, and key zeroization.			
	Secret and private keys established using manual methods may be entered or output in plaintext form.		Secret and private keys established using manual methods shall be entered or output encrypted or with split knowledge procedures	
EMI/EMC	47 CFR FCC Part 15. Subpart B, Class A (Business use). Applicable FCC requirements (for radio).		47 CFR FCC Part 15. Subpart B, Class B (Home use).	
Self-Tests	Power-up tests: cryptographic algorithm tests, software/firmware integrity tests, critical functions tests. Conditional tests.			
Design Assurance	Configuration management (CM). Secure installa ion and genera ion. Design and policy correspondence. Guidance documents.	CM system. Secure distribution. Functional specification.	High-level language implementation.	Formal model. Detailed explanations (informal proofs). Preconditions and post-conditions.
Mitigation of Other Attacks	Specification of mitigation of attacks for which no testable requirements are currently available.			

5.1.3 Pre-Validation List

The FIPS 140-1 and FIPS 140-2 Pre-Validation List is provided for information purposes only and is a voluntary list. Posting on the list does not guarantee a final FIPS 140-2 or FIPS 140-2 validation. The following phases describe the FIPS 140-1 and FIPS 140-2 pre-validation process. The status of each cryptographic module in the process is identified in the Pre-Validation List.

1. ***Implementation Under Test (IUT)***

    - There exists a viable contract between the vendor and Cryptographic Module Testing (CMT) laboratory for the testing of the cryptographic module.

    - The cryptographic module is resident at the CMT laboratory.

    - All of the required documentation is resident at the CMT laboratory. (Note: if the vendor requires the CMT lab personnel to test the cryptographic module on-site, all documents must be on-site with the module.)

2. ***Validation Review Pending***

    - A complete set of testing documents has been submitted to NIST and CSE for review. The set includes: a draft certificate, a summary description of the module, a detailed test report, a non-proprietary security policy, and website information. In addition, some CMT labs include a separate physical testing report.

- A signed letter from the laboratory stating its recommendation for validation has been received by NIST and CSE.

3. ***Validation Review***

    - NIST and CSE reviewers have been assigned.

    - NIST and CSE perform a preliminary review of the test documents (if required). NIST and CSE perform a review of the test documents.

    - Comments are coordinated by NIST and CSE reviewers and combined into a set of comments sent to the CMT laboratory.

4. ***Validation Coordination*** (this process may be iterative)

    - Comments have been received by the CMT laboratory from NIST and CSE for resolution.

    - Additional testing is performed (if required).

    - Additional documentation is obtained (if required).

    - Comments resolution is developed for resubmission to NIST and CSE.

    - Testing documents are updated for resubmission to NIST and CSE.

    - Responses to comments and revised test documents are submitted to NIST and CSE.

5. ***Validation Finalization***

    - Final resolution of validation review comments are submitted to NIST and CSE.

    - Testing documents are updated, based on resolutions and are submitted to NIST and CSE.

    - Certificate number is assigned.

    - Certificate printing and signature process is initiated.

5.1.4 Validated Modules List

The Validated Modules List includes the following information for each cryptographic module that has been validated against FIPS 140-1 or FIPS 140-2:

- Vendor Name and Point-of-Contact (POC)
- Module Name and Version Number
- Validated to FIPS 140-1 or 140-2
- Module Type (software, hardware, firmware)
- Date of Validation (and revalidation, if applicable)
- Level(s) of Validation
- A description of the Module including the validated algorithms
- Links to: Validation certificate, Security Policy, company web site, and the technical point of contact

Note that a module on the list may be a product, may be used in multiple products from that vendor, or may be used in another vendor's product(s).

5.1.5 Effective Use of FIPS 140-2

When implementing cryptography in a product/system:

- Examine FIPS 140-2. Consider the requirements in each area. Determine those requirements that specify the features that are desired. Determine those requirements (if any) that are specified in FIPS 140-2 that were not originally considered. Specify the appropriate level in each area of the standard based on the acceptable level of risk, organization mission, and identification of assets.

- Examine Annex A of FIPS 140-2 *(Approved Security Functions for FIPS 140-2)* to ensure that the cryptographic module employs an Approved algorithm and supports a mode of operation approved for the desired security service(s) (e.g., symmetric key, asymmetric key, message authentication and hashing).

- Examine Annex B of FIPS 140-2 (*Approved Protection Profiles for FIPS 140-2*) to determine if an approved protection profile applies to any operating system associated with the proposed cryptographic module.

- Compare the proposed cryptographic design or prospective product's specifications to the requirements of Annex C of FIPS 140-2 (*Approved Random Number Generators for FIPS 140-2*) to determine conformance to the standard.

- Compare the proposed cryptographic design or prospective product's specifications to the requirements of Annex D of FIPS 140-2 (*Approved Key Establishment Techniques for FIPS 140-2*) to determine conformance to the standard.

- Obtain or develop cryptographic modules that meet or exceed the selected levels and/or conform to an approved protection profile.

5.2 National Voluntary Laboratory Accreditation Program (NVLAP)

The NIST National Voluntary Laboratory Accreditation Program (NVLAP) accredits testing organizations, based on technical accreditation requirements and quality system requirements. NVLAP assesses the testing organization against the NVLAP accreditation requirements to determine if the organization is competent to perform specific tests and calibrations. Competence is defined as the ability of a laboratory to meet the NVLAP conditions and to conform to the criteria in NVLAP publications for specific calibration and test methods.

5.3 Industry and Standards Organizations

The next higher level of testing, above algorithm and module testing, is at the product level. Products are tested by the vendor, standards organizations, and by independent verification and validation (IV&V) organizations. Vendors test their products to ensure that they function properly and in a secure manner. Cryptographic modules and components may be integrated or embedded into these products. For government applications, the embedded cryptographic modules must meet the requirements of FIPS 140-2. At product level testing, the product should not compromise or circumvent the cryptographic features, resulting in a non-secure device. Products should be tested to the Common Criteria, Version 2.2.

5.3.1 National Information Assurance Partnership (NIAP)

The National Information Assurance Partnership (NIAP) is a U.S. Government initiative that was created to meet the security testing needs of both information technology (IT) consumers and producers. NIAP is a collaboration between the NIST and the National Security Agency (NSA). The partnership combines the extensive IT security experience of both agencies to promote the development of technically sound security requirements for IT products and systems and the development of appropriate measures for evaluating those products and systems.

The long-term goal of NIAP is to help increase the level of trust that consumers have in their information systems and networks through the use of cost-effective security testing, evaluation, and validation programs. In meeting this goal, NIAP seeks to:

- Promote the development and use of evaluated IT products and systems;

- Champion the development and use of national and international standards for IT security;

- Foster research and development in IT security requirements definition, test methods, tools, techniques, and assurance metrics;

- Support a framework for international recognition and acceptance of IT security testing and evaluation results; and

- Facilitate the development and growth of a commercial security testing industry within the U.S.

The focus of the Common Criteria Evaluation and Validation Scheme (CCEVS) is to establish a national program for the evaluation of IT products for conformance to the *International Common Criteria for Information Technology Security Evaluation (CC)*. The CC is a *voluntary* standard that is used to describe the security properties (functional and assurance) of IT products (or classes of products). These criteria are used throughout the international community in establishing security requirements for products. The CC also defines a Protection Profile (PP) construct that allows consumers or developers to create standardized sets of security requirements that will meet their needs. For example, PPs have been developed for firewalls, operating systems, and database management systems. IT products that are specified using the CC may then be evaluated for conformance to their CC specifications.

IT security evaluations are conducted by commercial testing laboratories that are accredited by NIST'S National Voluntary Laboratory Accreditation Program (NVLAP) and approved by the Validation Body. These approved testing laboratories are called Common Criteria Testing Laboratories (CCTL). NVLAP accreditation is one of the requirements for becoming a CCTL. The purpose of the NVLAP accreditation is to ensure that laboratories meet the requirements of ISO/IEC 17025, *General Requirement for the Competence of Calibration and Testing Laboratories* and the specific scheme requirements for IT security evaluation. Other requirements for CCTL approval are CCEVS-specific and are outlined in scheme policies and publications.

The Validation Body assesses the results of a security evaluation conducted by a CCTL within the scheme and, when appropriate, issues a CC certificate. The certificate, together with the validation report, confirms that an IT product or PP has been evaluated at an accredited laboratory using the Common Evaluation Methodology (CEM) for conformance to the CC. The certificate also confirms that the IT security evaluation has been conducted in accordance with the provisions of the scheme and that the conclusions of the testing laboratory are consistent with the evidence presented during the evaluation.

The Validation Body maintains a NIAP Validated Products List (VPL) containing all IT products and PPs that have successfully completed evaluation and validation under the scheme.

The CC and FIPS 140-2 are different in their abstractness and focus. The four security levels in FIPS 140-2 do not map directly to specific CC EALs or to CC functional requirements. A CC evaluation of a product containing cryptography does not supersede or replace a validation of a cryptographic module through the CMVP, and a CC certificate cannot be used to meet the mandatory requirements of FIPS 140-2.

5.3.2 Certification and Accreditation

The highest level of testing is at the application or system level within the operational environment. At a Federal agency, this level of testing is conducted during the certification phase. NIST SP 800-37, *Guide for the Security Certification and Accreditation of Federal Information Systems*, provides guidelines for certifying and accrediting information systems supporting the executive agencies of the federal government. These guidelines have been developed to:

- Enable more consistent, comparable, and repeatable evaluations of the security controls applied to Federal information systems;

- Promote a better understanding of enterprise-wide mission risks resulting from the operation of information systems;

- Create more complete, reliable, and trustworthy information for authorizing officials---- facilitating more informed security accreditation decisions; and

- Help achieve more secure information systems within the Federal government, including the critical infrastructure of the United States.

Certification is the comprehensive analysis of both the technical and non-technical security controls and other safeguards of a system. The security assessment conducted during the Certification Phase establishes the extent to which a particular system meets the security requirements for its mission and operational needs. Certification is performed in support of managements' authorization to operate a system. Certification examines the system in the operational environment and examines external systems that are networked to the system under test (SUT). One of the major tasks of certification is to verify that external systems are not able to compromise or circumvent the security features (including cryptographic features) of the SUT.

The official management decision to authorize processing (i.e. security accreditation), given by a senior agency official, is applicable to a particular environment of operation, and explicitly accepts the risk to agency operations, agency assets, or individuals after the implementation of an agreed-upon set of security controls. By accrediting the information system, the agency official is not only responsible for the security of the system, but is also accountable for adverse impacts to the agency if a breach of security occurs. Security accreditation provides a form of quality control.

At all levels of testing, it is important to be able to trace the implemented cryptographic controls and other security features through the requirements back to a standard.

Chapter 6

SELECTING CRYPTOGRAPHY - THE PROCESS

The process used to select cryptographic mechanisms is similar to the process used to select any IT mechanism. This selection process is documented in the system development life cycle (SDLC) model. Many SDLC models exist that can be used by an organization to effectively develop an information system. A traditional SDLC is a linear sequential model. This model assumes that the system will be delivered near the end of its development life cycle. Another SDLC model uses prototyping, which is often used to develop an understanding of system requirements without developing a final operational system. More complex models have been developed to address the evolving complexity of advanced and large information system designs. The SDLC model is embedded in any of the major system developmental approaches:

- *Waterfall* - the phases are executed sequentially;

- *Spiral* - the phases are executed sequentially, with feedback loops to previous phases;

- *Incremental development* - several partial deliverables are constructed, and each deliverable has incrementally more functionality. Builds are constructed in parallel, using available information from previous builds. The product is designed, implemented, integrated and tested as a series of *incremental builds*.

- *Evolutionary* - there is replanning at each phase in the life cycle, based on feedback. Each phase is divided into multiple project cycles with deliverable measurable results at the completion of each cycle.

Security should be incorporated into all phases, from initiation to disposition, of an SDLC model. The goal of the selection process is to specify and implement cryptographic methods that address specific agency/organization needs. Appendix E provides a list of standards, recommendations, guidelines, and assessment documents for Federal cryptographic mechanisms (with URLs for electronic versions).

Prior to selecting a cryptographic method, an agency should consider the operational environment, the application requirements, the types of services that can be provided by each type of cryptography, and the cryptographic objectives that must be met when selecting applicable products. Based on the requirements, several cryptographic methods may be required. For example, both symmetric and asymmetric cryptography may be needed in one system, each performing different functions (e.g., symmetric encryption, and asymmetric digital signature and key establishment).

The following high level questions should be addressed in determining the appropriate cryptographic mechanisms, policies, and procedures for a system:

- How critical is the system to the organization's mission, and what is the FIPS 199 impact level?

- What are the performance requirements for cryptographic mechanisms (e.g., communications throughput, processing latency)?

- What inter-system and intra-system compatibility and interoperability requirements need to be met by the system (e.g., algorithm, key establishment, and cryptographic and communications protocols)?

- What are the security/cryptographic objectives required by the system (e.g., content integrity protection, source authentication required, confidentiality, availability)?

- For what period of time will the information need to be protected?

- What regulations and policies are applicable in determining what is to be protected?

- Who selects the protection mechanisms that are to be implemented in the system?

- Are the users knowledgeable about cryptography, and how much training will they receive?

- What is the nature of the physical and procedural infrastructure for the protection of cryptographic material and information (e.g., storage, accounting and audit, logistics support)?

- What is the nature of the physical and procedural infrastructure for the protection of cryptographic material and information at the facilities of outside organizations with which cryptographically protected communications are required (e.g., facilities and procedures for protection of physical keying material)?

The answers to these questions can be used to formulate a development approach for integrating cryptographic methods into existing or new systems. A sound approach in integrating cryptographic methods is to develop requirements that are derived from the protection goals and policies for the system.

The following areas relate specifically to cryptography and should be included when developing requirements:

- Security of the cryptographic module[19]
- Hardware versus software implementation
- Applying cryptography in a networked environment
- Implementing Approved algorithms
- Symmetric key versus asymmetric key cryptography
- Key length
- Key management infrastructure
- Cryptographically-protected interoperation with external organizations (Federal, state, local, foreign, private sector)

It is important to be able to demonstrate *traceability* from the requirements back to the policies and goals and associated risk assessment.

There are other issues to be addressed in achieving overall security. Cryptography is best used when it is designed as an integrated part of the system, rather than as an add-on feature. When this cannot be done, cryptographic functions should be carefully added so that the security that they are intended to provide is not compromised. The least effective approach to implementing cryptography is to immediately begin implementing technical approaches. (Note: implementing technical solutions without determining the requirements is never recommended.) Cryptographic methods are intended to address specific security risks and threats. Therefore, implementing only cryptographic methods, and no other security mechanisms in a system, will not necessarily provide adequate security. A cryptographic solution may be initially implemented as a pilot project to ensure that the solution is effective.

In many cases, interoperability is required with organizations that already have an installed base of cryptographic controls. Where justified by a requirement's analysis or business case, *and* where compliance with Federal standards and recommendations can be maintained, cryptographic components that are interoperable with the installed base should be selected. Interoperability considerations include:

- Algorithm
- Key size
- Modes of operation

---

[19] The *Recommendation for Key Management*, Special Publication 800-57, provides specific criteria for the selection of algorithms and key lengths. FIPS 140-2 provides physical and logical cryptographic module security requirements.

- Key establishment protocols
- Communications protocols

Figure 9 illustrates the system development life cycle phases and the security-related tasks that are to be performed within each phase when specifying, selecting, and implementing cryptographic methods. These tasks should be performed when acquiring and implementing new systems requiring cryptographic products or when acquiring cryptographic products for existing systems. The tasks listed in Figure 9 and discussed in this guideline are tailored to cryptographic methods.

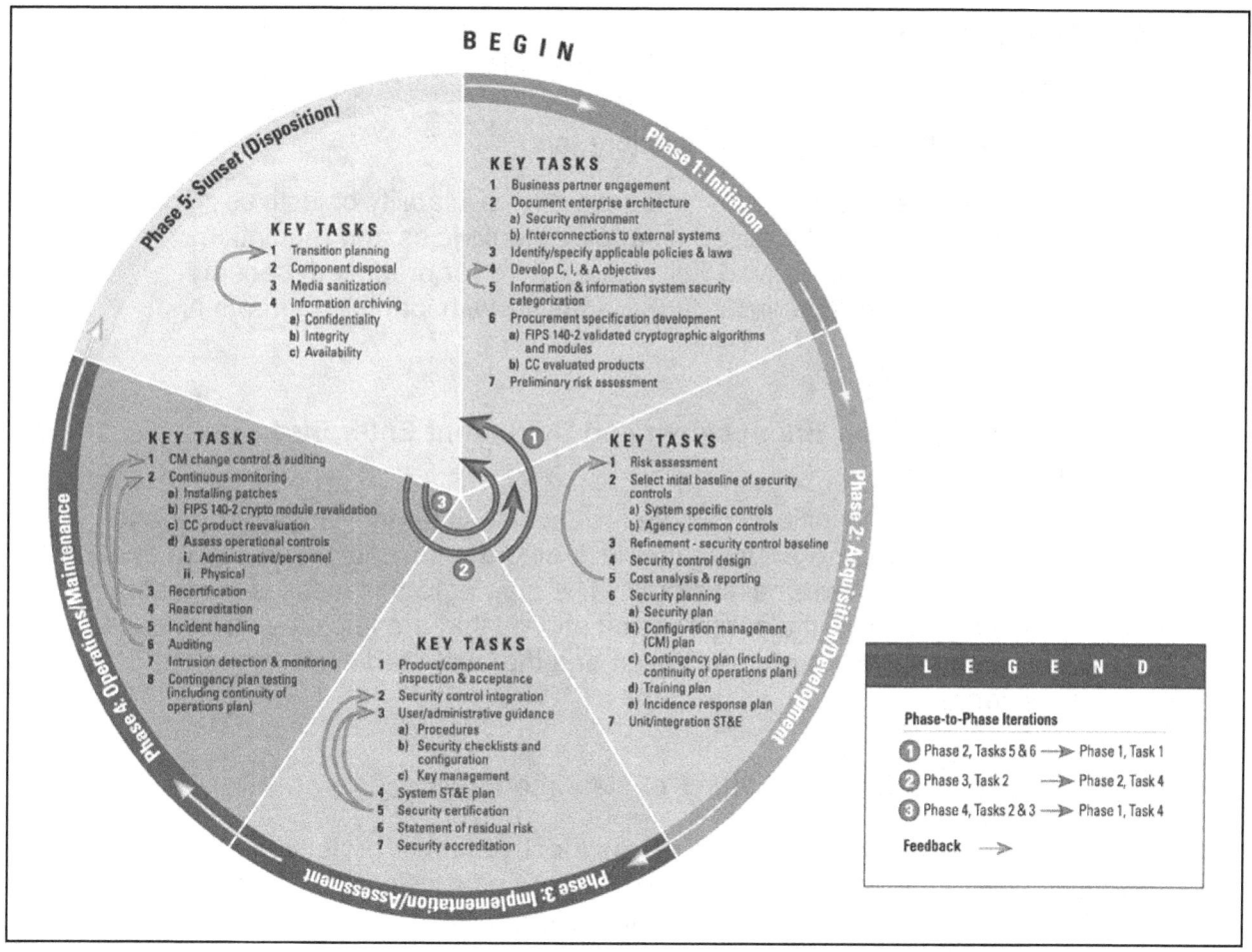

Figure 9: Information Security in the System Development Life Cycle

Also, because of the role that cryptographic controls play in protecting sensitive information, greater emphasis should be placed on developing applicable documentation, (e.g., user procedures and crypto-officer manuals) and implementing ongoing operational controls, (e.g., key management).

In general, the tasks in Figure 9 are listed in sequence; (e.g., select the initial base of security controls prior to performing security certification). Realistically, some of the tasks may be executed concurrently; for example, performing a risk assessment could be performed concurrently with developing objectives, and selecting and implementing controls could be performed concurrently with developing applicable documentation. To ensure that a cryptographic product is correctly implemented to provide appropriate security functionality, all tasks should be performed, particularly documentation development, training, and ongoing operations tasks. The SDLC phases and information security tasks are described in more detail in the following sections, with a focus on cryptography.

6.1 Phase 1: Initiation

The focus in Phase 1 is to:

- Document the enterprise architecture, and
- Document the confidentiality, integrity, and availability objectives. These objectives are partially based on applicable policies and regulations. Objectives are also derived from the existing (or proposed) security environment and a preliminary risk analysis with identified threats and vulnerabilities.

6.1.1 Business Partner Engagement and Document Enterprise Architecture

The first two tasks in the initiation phase are: business partner engagement and to document the Enterprise Architecture, including the security environment and any interconnections to external systems. The cryptographic focus is to identify the Approved cryptographic algorithms and modes that are implemented in each information system within the organization and that are used in interconnected external systems.

6.1.2 Identify/Specify Applicable Policies and Laws

IT security policy helps establish standards for IT resource protection by assigning program management responsibilities and providing basic rules, guidelines, and definitions for everyone in the organization. Policy helps prevent inconsistencies that can introduce risks, and serves as a basis for the enforcement of more detailed rules and procedures. IT security policy is generally formulated with input from many members of an organization, including security officials, users, managers, and IT resource specialists.

After policies are established, requirements (including security and cryptographic requirements) are specified and an overall system design is developed. The system design includes software and hardware implementations, procedures, environmental requirements, physical security considerations, etc.

Policies (and applicable laws and regulations) can be used effectively in the design, development and implementation of cryptographically-based controls and procedures, if they are implemented in a practical (real-world) manner.

The following are some topics that should be addressed when developing cryptography policies and requirements:

1. Policies regarding algorithm usage and algorithm parameter configuration (e.g., key sizes),

2. Policies regarding the classes of users (e.g., crypto-officers, networked users, operators) that may use the cryptographic methods and for assigning associated privileges,

3. Identification and authentication requirements when a user initially accesses a system or cryptographic module,

4. Procedures employed when adding, modifying, or deleting users and user privileges associated with cryptographic methods/products,

5. Policies defining when confidentiality controls, integrity controls, and advanced authentication techniques are required,

6. Security measures relating to the physical environment of a cryptographic method/product,

7. Audit procedures,

8. Guidelines for requiring non-repudiation,

9. Guidelines for performing risk assessments to:

    a. Ensure that the unique risks of an IT system are considered,

    b. Evaluate the potential risks and determine the level of control required to minimize the risks, commensurate with the cost or value of the data,

10. Key Management policies, including key distribution, generation, use, destruction, and archiving,

11. Backward compatibility of software/hardware and architecture,

12. Forward compatibility with envisioned future developments, such as new cryptographic techniques, digital signature systems, authentication mechanisms, FIPS, implementation recommendations and guidelines, and

13. Interoperability among governments, commercial communities, law enforcement communities, etc.

6.1.3 Develop C, I, and A Objectives

The fourth task in the initiation phase is to develop confidentiality, integrity, and availability objectives. These objectives are at a high level and should address security, in general, and cryptography, specifically. Example 1 lists sample security objectives.

Example 1: Security Objectives

1. *Integrity*: The correctness of cryptographic keys and other critical security parameters must be preserved. Authentication, authorization and non-repudiation should be supported. The correctness of the security mechanisms/features should be ensured.

2. *Availability*: The security mechanisms/features should be continually available (at least 99.5% of the time). Availability periods must be tailored to particular systems or environments. Response time to suspected compromise (for example, disclosure or modification) should be minimized. Systems should be responsive and adaptable to changing security requirements and threats.

3. *Integrity and Authentication:* Digital signatures may be used to validate the:

    - Identity of the signer of a message and
    - Integrity of the information received from the signer of that information.

6.1.4 Information and Information System Security Categorization and Procurement Specification Development

Task 5 is *Information and Information System Security Categorization*, and task 6 is *Procurement Specification Development*, including specifying FIPS 140-2 validated cryptographic algorithms and modules and CC evaluated products.

FIPS 199, *Standards for Security Categorization of Federal Information and Information Systems* and *Guide for Mapping Types of Information and Information Systems to Security Categories* (NIST SP 800-60) provide guidelines for assessing the security impact, or magnitude of harm, that can be expected to result from confidentiality and integrity compromises of various types of Federal information and information systems. The impact determinations can then be used with *Recommended Security Controls for Federal Information Systems* (NIST SP 800-53) to help select security controls and as an input to the risk assessment process. Cryptographic security goals and policies are ultimately derived from a risk assessment.[20]

In the procurement specification development task, the goal is to develop the requirements/specifications for the proposed cryptographic methods. After the requirements have been developed, general selection criteria based on these requirements are produced. Finally, categories of methods that meet the requirements are identified. The security requirements are based on user needs and estimates of an organization's resources to meet proposed requirements. Requirements should be detailed - this aids in product selection, implementation, and testing.

6.1.5 Cryptographic Method Example

Examples 2 and 3 focus on a specific cryptographic method, digital signatures, and illustrates how requirements may be derived from a high-level digital signature policy statement.

Example 2: Digital Signature Policy

Background: Historically, handwritten signatures were used to provide authenticity and liability for a document. The proposed successor to handwritten signatures is digital signatures.

Policy Statement. Digital signatures will be accepted as valid only if the user who verifies a signature has an acceptable level of assurance of the integrity of the electronic document that was signed and the identity of the signer of that document. In addition, the verifier must be able to trust that the signer will be held legally responsible for the information content of the document.

One of the digital signature policies is to ensure the integrity of electronic documents and provide non-repudiation of document origin. The requirements resulting from this policy include all three types: functional, assurance and environmental.

---

[20] NIST SP 800-30, *Risk Management Guide for Information Technology Systems*, provides guidelines for conducting risk assessments.

Example 3: Digital Signature Requirements

Requirements:

1. *Document preservation.* Associated signatures and the certificates necessary to verify those signatures must accompany electronic documents. A record of certificate validity must also be kept, along with an audit trail of document movement. Expert testimony about this entire procedure and the audit data collected will lay the foundation for the testimony if documents are required as evidence.

2. Digital signatures do not, by themselves, provide time-related information. A trusted time stamp is required to prove when a document was originated or received. This service must be provided by a trusted third party, who may be serving the purpose of a notary. The originator will generate a hash value for the document and send a copy of the document and the hash value to the trusted party that is serving as a notary. This trusted party could time and date stamp the hash value for the document, store a copy of the hash value and the document, keep an audit log of the action, and serve as an intermediary between the document's originator and receiver.

3. *Establishment of user and CA responsibility.* The document signer must be responsible for protecting the private key used to sign a document and obtaining time stamped document receipts, when required. A document verifier must ensure that all certificates used to verify a received document are valid at the time the document is received, and ensure that the received document is time stamped and the required information is archived in case of litigation. The CA is responsible for protecting the private key used to sign certificates, establishing the identity of its subscribers, and providing certificates and revocation information in a timely manner.

4. Each entity, whether an originating or sending entity or a CA, must maintain an audit log of digital signature related activity, including messages sent and received, activity by persons associated with the signature process and other security-relevant events.

5. Policies and procedures must be established to ensure that control is maintained on all processes involving the electronic authorization and authentication of electronic documents.

6. Policies and procedures must be established that will ensure that an approved process protects the distribution and communication of authorizations.

Table 4 identifies security objectives and requirements for cryptographic components that may be addressed by cryptographic controls. The purpose of the table is to provide individuals with a *roadmap* to identifying cryptographic requirements in families that will meet the needs of a system in an organization. The security control families are documented in NIST SP 800-53, *Recommended*

Security Controls for Federal Information Systems. The security controls are organized into *families* for ease of use in the control selection and specification process. Each family contains security controls related to the security function of the family. After the families are selected, the specific requirements are extracted from NIST SP 800-53. The specific requirements may need to be refined/tailored to meet security objectives. Table 4 is not intended to list all the specific requirements; rather it serves as a reference guide to the security control families. To effectively use the table, it is important to have documented the objectives that must be addressed. These objectives were developed in tasks in the Initiation Phase.

- *Column 1, Cryptographic Area,* covers the areas related to the design and implementation of a cryptographic product/module. Some examples are roles and services, physical security, and cryptographic key management.

- *Column 2, Security Control Families,* list the requirements that address the security objectives for a cryptographic category. The requirements are listed by *NIST SP 800-53* for example, Audit and Accountability, Awareness and Training, Configuration Management, Contingency Planning, Identification and Authentication, System and Services Acquisition, and System and Information Integrity.

- *Column 3* contains *Procurement Recommendations* to ensure the cryptographic requirements are adequately addressed.

Table 4: Cryptographic Area and Security Control Families		
Cryptographic Area	**Security Control Families**	**Procurement Recommendations**
Cryptographic Module Specification: cryptographic boundary; diagram configuration; security policies; *Cryptographic Algorithms:* Approved algorithms	*Security Planning* (PL), *System and Services Acquisition* (SA)	Verify the module or product[21] is on the FIPS 140-2 validated modules list. No unique requirements beyond specifying the required algorithms and modes.
Cryptographic Module Ports and Interfaces: physical and ogical data paths	*Certification, Accreditation, and Security Assessments* (CA), *System and Services Acquisition* (SA), *System and Information Integrity*	The requirements for the cryptographic module should be consistent with the requirements for the other components of the

---

[21] A validated cryptographic module may be embedded in a product that is submitted for validation.

Table 4: Cryptographic Area and Security Control Families		
Cryptographic Area	**Security Control Families**	**Procurement Recommendations**
	(SI), System and Communications Protection (SP).	system.
Roles, Services, and Authentication: roles and associated services; authorization and access control mechanisms	Access Control (AC), Audit and Accountability (AU), Identification and Authentication (IA), Systems and Services Acquisition (SA), System and Communications Protection (SP)	A system administrator may include the cryptographic officer role.
Physical Security: physical security configuration and mechanisms; specify features or testing procedures. (Includes EFP/EFT).	System Maintenance (MA), Media Protection (MP), Physical and Environmental Protection (PE), Systems and Services Acquisition (SA)	Verify that the physical controls adequately protect the cryptographic module.
Operational Environment: access, authorization, audit controls; identify critical security parameters and cryptographic data	Referenced PPs evaluated at EAL2, EAL3+, or EAL4	Include applicable cryptographic module events in the auditable events.
Cryptographic Key Management: random number generation, key generation, key establishment, key entry and output, key storage, and key destruction	System and Communications Protection (SP)	Include the key establishment and key management procedures in the administrative guidance and user responsibilities in the user guidance.
Self-Tests: identify power-up and conditional tests	System and Information Integrity (SI)	No unique requirements beyond specifying the tests for the Approved algorithms.
Design Assurance: describe the design of the software/hardware/firmware; correspondence between	Configuration Management (CM), Security Planning (PL), System and Services Acquisition (SA),	The use of CM tools and life cycle support products should be the same for the cryptographic module and

Table 4: Cryptographic Area and Security Control Families		
Cryptographic Area	**Security Control Families**	**Procurement Recommendations**
the design and the security policy	*System and Information Integrity (SI)*	the system
EMI/EMC[22]: FCC conformance requirements	EMI/EMC FCC part 15, Subpart B, Class A (business use) or Class B (home use) requirements	No unique requirements beyond identifying the FCC requirements.
Mitigation of Other Attacks	*Security Planning (PL), System and Services Acquisition (SA)*	*Perform a vulnerability analysis.*

6.1.6 Preliminary Risk Assessment

The final key task in the initiation phase is to perform a preliminary risk assessment and, specifically, identify the unique requirements associated with each information system. After the risk assessment has been performed, policies should be developed regarding the use of evaluated operating systems and validated cryptographic modules in a range of environments. Policies that have been previously written may need to be revised or tailored throughout the SDLC.

Risk management consists of two components:

- *Assessing* risks using a risk-based approach to determine the impact of given losses and the probability that these losses will occur. The major losses addressed by cryptographic methods are the unauthorized disclosure and modification of data.

- *Selection and implementation* of countermeasures that either reduce the probability of threat occurrence or minimize the impact of loss. The goal is to reduce the risk to an acceptable level.

The purpose of an IT risk management process is to ensure that the impacts of threats are known and that cost-effective countermeasures are applied to determine *adequate security* for a system. *Adequate security* is defined in OMB Circular A-130, Appendix III, as "security commensurate with the risk and magnitude of harm resulting from the loss, misuse, or unauthorized access to or modification of information. This includes assuring that systems and applications used by an agency operate effectively and provide appropriate confidentiality,

---

[22] Electromagnetic Interference/Electromagnetic Compatibility

integrity, and availability through the use of cost-effective management, personnel, operational, and technical controls." This definition explicitly emphasizes the risk-based policy for cost-effective security established by the Federal Information Security Management Act (FISMA).

Risk assessment, the process of analyzing and interpreting risk, includes the following activities.

- System characterization
    - Identify assets
    - Assess current security and protection mechanisms
- Identify and classify threats affecting:
    - Integrity
    - Confidentiality
    - Availability
- Identify potential losses (likelihood determination and impact analysis)
    - Classify potential losses by criticality and sensitivity
- Identify potential controls
    - Evaluate potential countermeasures so that implementation decisions can be made
    - Perform cost/benefit analysis for proposed controls. (The analysis should include both *monetary* and *non-monetary* perspectives.)

The following example illustrates how cryptographic methods can address integrity and non-repudiation threats.

Example 4: Threat Mitigation

Security Control to Mitigate Threat to Integrity: Both secret key and public key cryptography can be used to ensure integrity. When secret key cryptography is used, a message authentication code (MAC) is generated. Typically, a MAC is stored or transmitted with the data. When the integrity of the data is to be verified, the MAC is generated on the current data and compared with the previously generated MAC. If the two values are equal, the integrity (i.e., authenticity) of the data is verified.

In public key cryptography, a secure hash algorithm is used to create a message digest. The hash will change if the message is modified. The hash is then signed with a private key. The hash may be stored or transmitted with the data. When the integrity of the data is verified, the hash is recalculated and the corresponding public key is used to verify the integrity of the message.

> **Security Control to Mitigate Threat to Non-repudiation:** Data is digitally signed by applying the originator's private key to the data. The resulting digital signature can be stored or transmitted with the data. Any party using the public key of the signer can verify the signature. If the signature is verified, then the verifier has confidence that the data was not modified after being signed and that the owner of the public key was the signer. A certificate *binds* the public key to the identity of the signer.

A risk assessment is performed for all new and existing systems, even if it is not called a formal risk assessment. The type of risk assessment that is performed is usually a qualitative analysis, rather than a formal quantitative analysis, and the results are used in developing the system requirements and specifications. A team composed of users, system developers, and security specialists typically conducts the risk assessment. The scope of this task varies depending on the sensitivity of the information and the number and types of risks that need to be addressed. For systems with minimal security requirements, the risk assessment may be a few pages in length.

6.2 Phase 2: Acquisition/Development

The first task in the acquisition/development phase is to update the risk assessment that was performed in phase 1. After completing the update, the next task is to select the initial baseline of security controls from SP 800-53. This initial baseline is reviewed and revised based on the risk assessment.

6.2.1 Selecting Cryptographic Controls

The second task in the acquisition/development phase is to identify categories of cryptographic methods/techniques that meet the requirements and mitigate the specific risks. There may be more than one method category that can mitigate each risk. For example, both MACs and digital signatures can protect against the undetected modification of data.) For many of the methods, there are assurance features that increase the confidence that the method performs correctly.

Section 5.1.5 of this guideline, "Effective Use of FIPS 140-2," lists a number of steps that may prove helpful in the selection of cryptographic countermeasures.

Table 5 lists the technical and assurance features that meet the technical and assurance requirements documented in Table 4. The features in Table 5 map directly to the requirements listed in Table 4.

- *Column 1* lists the Cryptographic Area.
- *Column 2* identifies the risks that apply to a cryptographic category, for example, unauthorized access or unauthorized disclosure.
- *Column 3* lists the technical and assurance requirements that are applicable to a cryptographic category and mitigate the potential risks.

December 2005   Implementing Cryptography

       Where applicable, the requirements are numbered and listed in ascending order of protection, to address increasing levels of risk. The levels vary from 1 to 4, corresponding to the security levels in FIPS 140-2.

- *Column 4* lists the FIPS and SPs that describe the features. The information included in the cryptographic category columns is the same as that presented in the requirements table (Table 3). This provides for traceability from the requirements to the methods and features.

Tables 4 and 5 illustrate traversing from a high level of abstraction in the requirements to a lower level of granularity in identifying specific features. It is important to understand that Table 4 does **not** specify the necessary conditions for the secure implementation of a product in a particular system/application. This task is left to those who implement the system.

Table 5: Risks and Cryptographic Features			
Cryptographic Area	**Risks**	**Technical and Assurance Requirements**	**Cryptographic Toolkit Reference**
<u>Cryptographic Module Specification</u>: Specify cryptographic boundary; specify cryptographic algorithms; diagram configuration; specify security policy; describe operational and error states.	Incorrect implementation	Cryptographic requirements addressed in overall system/product requirements. Security policy (including security rules), configuration block diagram.	
<u>Cryptographic Algorithms</u> (identify FIPS-approved algorithms and other cryptographic algorithms): Encryption	1. Unauthorized disclosure of data or undetected modification of data (intentional and accidental) during transmission or while in storage. 2. Denial of service. 3. Session capture. 4. Man-in-the-	1. FIPS-approved AES algorithm or three key TDEA algorithm; conformance tests.	FIPS 197: *AES* NIST SP 800-67: *Recommendation for the Triple Data Encryption Algorithm (TDEA) Block Cipher.*

Table 5: Risks and Cryptographic Features			
Cryptographic Area	**Risks**	**Technical and Assurance Requirements**	**Cryptographic Toolkit Reference**
	middle attack.		
Cryptographic Algorithms: Block Cipher Modes of Operation	(same as above)		DRAFT NIST SP 800-38, Recommendation for Block Cipher Modes of Operation:
Cryptographic Algorithms: Cryptographic Modules	(same as above)	FIPS-approved cryptographic algorithms; conformance testing.	FIPS 140-2: Security Requirements for Cryptographic Modules.
Cryptographic Algorithms: Hash Functions	(same as above)	Secure Hash Algorithm, message digest; conformance tests.	FIPS 180-2: Secure Hash Standard.
Cryptographic Algorithms: Digital Signatures.	(same as above)	Digital Signature Algorithm (DSA), RSA, ECDSA, digital signature generation/verification; message digest; random/pseudorandom number generation; hash function. Algorithms for generating primes p and q; private key generation; conformance tests. Cryptographic requirements addressed in overall system/product requirements.	FIPS 186-3: Digital Signature Standard.
Cryptographic Algorithms: Random Number Generation.		Algorithms for generating deterministic random bit generators; conformance tests.	Draft SP 800-90, Recommendation for Random Number Generation Using Deterministic Random Bit Generators.

Table 5: Risks and Cryptographic Features

Cryptographic Area	Risks	Technical and Assurance Requirements	Cryptographic Toolkit Reference
		Cryptographic requirements addressed in cryptographic algorithm and application documents.	Generators.
Cryptographic Module Ports and Interfaces: physical and logical input and output data paths.	Unintentional output of plaintext data. Design error.	Physical/logical separation of data input /output ports, control input, status output, data input, data output; documentation of the interfaces and input and output data paths.	FIPS 140-2: *Security Requirements for Cryptographic Modules.*
Roles, Services, and Authentication: Roles and associated services; authorization and access control mechanisms.	1. Unauthorized access by authorized/ unauthorized individuals. 2. Masquerade. 3. Password compromise. 4. Replay attacks.	1. Role-based authentication mechanisms. 2. Identity-based authentication mechanisms, maintenance-access interface; documentation of the authorized roles, services, operations, and functions	FIPS 140-2: *Security Requirements for Cryptographic Modules.*
Roles, Services, and Authentication :(continued)	(same as above)	1. Token based authentication. 2. Biometrics based authentication. 3. Cryptographic authentication protocols (secret key and public key cryptosystems).	FIPS 190: *Advanced Authentication.* NIST SP 800-38B and C, *Recommendation for Block Cipher Modes of Operation.*
Roles, Services, and Authentication: (concluded)	(same as above)	1. Digital signature algorithm.	FIPS 196: *Entity Authentication Using Public Key Cryptography.*

Table 5: Risks and Cryptographic Features			
Cryptographic Area	**Risks**	**Technical and Assurance Requirements**	**Cryptographic Toolkit Reference**
		2. Digital signatures.	
		3. Random/ pseudorandom number generator.	
		4. Unilateral authentication protocol.	
		5. Mutual authentication protocol.	
		Cryptographic requirements addressed in overall system/product requirements.	
Physical Security: Specify physical security configuration and mechanisms; specify features or testing procedures. (Includes *EFP/EFT*).	1. Unauthorized physical access to the contents. 2. Unauthorized use or modification, e.g., module substitution. 3. Unusual environmental conditions or fluctuations that results in disclosures of critical security parameters. 4. Unauthorized disclosure of plaintext critical security parameters.	1. Production grade enclosures. 2. Tamper evidence, or tamper resistance. 3, 4. Tamper response of shutdown of the module; zeroization of plaintext security keys and other unprotected critical security parameters (CSPs). 1, 2, 3. Specification of the physical embodiment, description of the applicable physical security mechanisms. 4. Specification of the environmental failure protection features, documentation of the environmental failure tests performed and the results.	FIPS 140-2: *Security Requirements for Cryptographic Modules.*

Table 5: Risks and Cryptographic Features

Cryptographic Area	Risks	Technical and Assurance Requirements	Cryptographic Toolkit Reference
Operational Environment: Specify access, authorization, audit controls; identify critical security parameters (CSPs) and cryptographic data.	1. Unauthorized access by authorized/ unauthorized individuals 2. Undetected modification of cryptographic component 3. Unauthorized modification, substitution, insertion, and deletion of cryptographic keys and other CSPs.	Level 1: Single operator, executable code, approved integrity technique. Level 2: Referenced PPs evaluated at EAL2 with specified discretionary access control mechanisms and auditing. Level 3: Referenced PPs plus trusted path evaluated at EAL3 plus security policy modeling. Level 4: Referenced PPs plus trusted path evaluated at EAL4.	FIPS 140-2: *Security Requirements for Cryptographic Modules.*
Cryptographic Key Management: Specify random number generation, key generation, key establishment, key entry and output, key storage, and key destruction.	1. Unauthorized disclosure, modification, and substitution of secret/private keys. 2. Unauthorized substitution and modification of public keys.	Key entry/output: Levels 1, 2. plaintext. Levels 3, 4. encrypted keys or split knowledge for manual-distribution. Key destruction: Zeroize all plaintext cryptographic keys and other unprotected CSPs. Specification of the FIPS-approved key generation algorithm; documentation of the key distribution techniques.	FIPS 140-2: *Security Requirements for Cryptographic Modules.* NIST SP 800-57: *Recommendation for Key Management.*
Cryptographic Key Management (concluded)	(same as above)	1. NIST-approved key generation algorithms. 2. Use of error detection code (message authentication code).	NIST SP 800-57: *Recommendation for Key Management.* NIST SP 800-56 *Recommendation on Key Establishment Schemes.*

Table 5: Risks and Cryptographic Features

Cryptographic Area	Risks	Technical and Assurance Requirements	Cryptographic Toolkit Reference
		1. Encrypted IVs. 2. Key naming. 3. Key encrypting key pairs. 4. Random number generation. Cryptographic requirements addressed in overall system/product requirements.	NIST SP 800-15, *Minimum Interoperability Specification for PKI Components (MISPC)*
<u>EMI/EMC</u>: identify FCC conformance requirements	Emanations	Conformance to FCC requirements. Cryptographic requirements addressed in overall system/product requirements.	FIPS 140-2: *Security Requirements for Cryptographic Modules.*
<u>Self-Tests</u>: Identify power-up and conditional tests.	1. Module malfunction. 2. Unauthorized disclosure of sensitive data.	Cryptographic requirements addressed in overall system/product requirements. Documentation on error conditions and actions to clear the errors; 1. Cryptographic algorithm test. 2. Critical functions test. 3. Pair-wise consistency test (for public and private keys). 5. Software/firmware	FIPS 140-2: *Security Requirements for Cryptographic Modules.*

Table 5: Risks and Cryptographic Features

Cryptographic Area	Risks	Technical and Assurance Requirements	Cryptographic Toolkit Reference
		load test. 6. Manual key entry test.	
<u>Design Assurance</u>: Describe the design of the software/hardware/ firmware; explain the correspondence between the design and the security policy.	Incorrect/invalid operation of the module.	Cryptographic requirements addressed in overall system/product requirements. Level 4. Formal model, informal proof.	FIPS 140-2: *Security Requirements for Cryptographic Modules.*
<u>Mitigation of Other Attacks</u>	Key compromise	Documented, if implemented.	FIPS 140-2: *Security Requirements for Cryptographic Modules.*

To clarify how all this information fits together, Example 5 walks through the two tables and illustrates the process of defining requirements, identifying risks, and then selecting cryptographic methods that meet those requirements and mitigates the risks. Additional explanatory information is included in brackets.

Example 5: Using Tables 4 and 5

Risk: Unauthorized disclosure of data or undetected modification of data (intentional and accidental) during transmission or while in storage [the risk was identified as the result of a risk assessment].

Security Controls: Implement FIPS-approved security methods for data integrity [the security control addresses the risk]. Tests [the cryptographic algorithm must be tested to ensure that it is compliant with the FIPS standard. Also, tests may be executed to ensure the algorithm was implemented correctly.].

Cryptographic Area: Cryptographic algorithms [these methods provide features that track any change, e.g., modification, insertion, deletion, to security-relevant data].

Technical and Assurance Requirements: FIPS-approved AES algorithm [implementations of the algorithm that have been tested and validated by NIST are compliant with the standard] NIST conformance tests [the tests are used to

> validate compliance with the standard].
>
> *Cryptographic Toolkit Reference:* FIPS 197: AES [specific AES modes can be used to calculate a data authentication code that provides for data integrity].
>
> *Procurement Recommendations*: Federal agencies that use cryptography to protect sensitive information must use systems that have been FIPS 140-2 validated.

When the cryptographic product/module is selected that meets the documented requirements, the product is then configured and tested. There are several types of testing that may be required, such as validation against FIPS 140-2, unit testing, and integration testing. Extensive testing of cryptographic controls is particularly important because of its role in ensuring the security of the overall system.

A second major component in the Acquisition Phase is to develop plans for users and cryptographic officers to inform them of their responsibilities in maintaining a secure system. Some of the plans are: security plan, configuration management plan, and training plan.

6.3   Phase 3: Implementation/Assessment

In the Implementation/Assessment phase, the focus is on configuring the system for use in the operational environment. This includes configuring the cryptographic controls. After the system has been configured, certification testing is performed to ensure that the system functions as specified and that the security controls are operating effectively.

The security provided by a cryptographic control depends on the mathematical soundness of the algorithm, the length of the cryptographic keys, key management, and mode of operation. A weakness in any one of these components may result in a weakness or compromise to the security of the cryptographic control. A weakness may be introduced at any phase of the system life cycle.

During product acquisition and development, it is the responsibility of the manufacturer of a cryptographic product to build a module that meets specified security requirements and conforms to a FIPS. However, conformance to a standard does NOT guarantee that a particular product is secure. To provide a level of assurance that the cryptographic product is secure, the product should be validated in the CMVP. The level of security in a cryptographic product/module must also be considered in the product selection phase. During this phase:

- Identify information resources and determine the sensitivity to and potential impact of losses. Determine security requirements based on risk assessment and applicable organizational security policies. Look at data sensitivity and the environment in which the data is placed. Consider threats to the data or application as a whole, and what level of risk is acceptable.

- Determine the acceptable safeguards for the system. Determine which cryptographic services provide an acceptable safeguard. Define those security features that are desirable for use and determine the appropriate security level from FIPS 140-2.

Finally, it is the responsibility of the integrator to configure and maintain the cryptographic module to ensure its secure operation (including maintenance of security, configuration management, and training plans). The use of a cryptographic product that conforms to a standard in an overall system does not guarantee the security of the cryptographic module or of the overall system. To summarize, the proper functioning of cryptography requires the secure design, implementation, and use of a validated cryptographic module.

There are many interdependencies among cryptography and other security controls, for example:

- *Physical Access Control.* Physical protection of a cryptographic module is required to prevent, or detect, physical replacement or modification of the cryptographic system and the keys within the system.

- *Logical Access Control.* Cryptographic modules may be embedded within a host system. With an embedded module, the hardware, operating system, and cryptographic software may be included within the cryptographic module boundary. Logical access control may provide a means of isolating the cryptographic software.

- *User Authentication.* Cryptographic authentication techniques may be used to provide stronger authentication of users. (Advanced authentication techniques are discussed in a later section.)

- *Assurance.* Assurance that a cryptographic module is properly and securely implemented is essential. The NIST CMVP provides assurance that a module meets stated standards.

- *Integrity Controls.* Cryptography may provide methods that protect security-relevant software, including audit trails, from undetected modification.

The major rule is: BUYER BEWARE!! Example 6 illustrates how important it is to correctly implement and manage all of the security and cryptography controls to ensure that keys are not compromised.

Example 6: Implementation Problems

1. Cryptographic algorithm may be strong, but the random number generator (RNG) may be weak.

> 2. RNG may be strong, but the Key Management may be weak.
> 3. Key Management may be strong, but the user authentication may be weak.
> 4. Authentication may be strong, but the physical security may be weak.

The following three rules guide the implementation of cryptography.

Determine what information must be protected using a cryptographic function.

The implementer should be aware of the information that is being cryptographically protected. Fields containing sensitive data should be identified, and then a determination should be made of what cryptographic functions should be applied to those fields: integrity, authenticity, and/or confidentiality.

Protect data prior to signature generation/verification and encryption/decryption. Be careful of how data is handled during these processes!

Implementers should be careful about how data is handled before it is encrypted and signed/verified. If data is stored in a central database and transferred to the computer only at the time the cryptographic function is performed, the data should be very carefully protected during transmission. If data is not carefully protected, an intruder could potentially alter data before a signature is generated, without the signer's knowledge. The data should be signed on the *signer's* machine, not in the central database.

Provide the capability for users to locally view all data that is being signed/encrypted.

Users should be able to see all the data that is being signed, and it should be clearly marked for the signer. Also, users should know what is encrypted. Not all data that is signed/encrypted should appear on one screen, but the user should be able to view all of the data before performing the cryptographic function.

6.4 Phase 4: Operations and Maintenance

In the Operations and Maintenance Phase, the goal is to ensure the continued secure operation of the cryptographic methods. One critical area is the life cycle management of cryptographic components.

The maintenance of cryptographic components is critical to ensure the secure operation and availability of the module/product. For example, cryptographic keys that are never changed, even when disgruntled employees leave, are not secure. The following are maintenance areas that need to be considered:

1. *Hardware/firmware* (e.g., new capabilities, expansion of the system to accommodate more users, replacement of non-functional equipment, change of platforms, hardware component upgrades, etc.)

2. *Software maintenance/update* (e.g., new capabilities, fixing errors, improved performance, key replacement, etc.)

3. *Application maintenance* (e.g., changes in roles and responsibilities, remote updates, updating passwords, deleting users from access lists, etc.)

4. *Key maintenance* (e.g., key archiving, key destruction, key change, etc.)

5. *Maintenance personnel.* Who is allowed to perform maintenance? Do maintenance personnel require clearances, or do authorized users monitor maintenance activities? What must be removed from the system prior to maintenance? How is the correctness of the maintenance procedure ascertained?

Configuration management (CM) is needed for areas 1 and 2. CM ensures the integrity of the management of system and security features through the control of changes made to a system's hardware, firmware, software, and documentation. The documentation may include user guidance, tests, test scripts and test documentation.

6.5 Phase 5: Sunset (Disposition)

When a system is shut down or transitioned to a new system, one of the primary responsibilities is ensuring that cryptographic keys are properly destroyed or archived. Long-term symmetric keys may need to be archived to ensure that they are available in the future to decrypt data. Signing keys use by CAs may also need to be maintained for signature verification. An individual's signing keys should not be archived. See SP 800-57 for further information.

December 2005 — Implementing Cryptography

APPENDIX A
ACRONYMS

AC	Access Control
AES	Advanced Encryption Standard
ANSI	American National Standards Institute
AU	Audit and Accountability
CA	Certification Authority
C&A	Certification and Accreditation
CC	Common Criteria
CCEVS	Common Criteria Evaluation and Validation Scheme
CCTL	Common Criteria Test Laboratory
CEM	Common Evaluation Methodology
CM	Configuration Management
CMIC	Certificate Issuing and Management Components
CMT	Cryptographic Module Testing
CMV	Cryptographic Module Validation
CMVP	Cryptographic Module Validation Program
CPS	Certification Practice Statement
CRL	Certificate Revocation List
CSE	Communications Security Establishment
CSP	Critical Security Parameters
DES	Data Encryption Standard
DRBG	Deterministic Random Bit Generator
DSA	Digital Signature Algorithm
DSS	Digital Signature Standard
DTR	Derived Test Requirement
EAL	Common Criteria Assurance Level
EC	Elliptic Curve
ECDSA	Elliptic Curve Digital Signature Algorithm

EMC	Electromagnetic Compatibility
EMI	Electromagnetic Interference
FBCA	Federal Bridge Certificate Authority
FCC	Federal Communications Commission
FIPS	Federal Information Processing Standard
FISMA	Federal Information Systems Management Act
HMAC	Keyed Hash Message Authentication Code
IEEE	Institute of Electrical and Electronics Engineers
IETF	Internet Engineering Task Force
ISO	International Organization for Standardization
IT	Information Technology
ITL	Information Technology Laboratory
IUT	Implementation Under Test
IV	Initialization Vector
IV&V	Independent Verification and Validation
MA	System Maintenance
MAC	Message Authentication Code
MISPC	Minimum Interoperability Specification for PKI Components
MP	Media Protection
NDRBG	Non-Deterministic Random Bit Generator
NIAP	National Information Assurance Partnership
NIST	National Institute of Standards and Technology
NVLAP	National Voluntary Laboratory Accreditation Program
NSA	National Security Agency
OCSP	Online Certificate Status Protocol
OMB	Office of Management and Budget
PE	Physical and Environmental Protection
PIN	Personal Identification Number
PKI	Public Key Infrastructure

P.L.	Public Law	
PP	Protection Profile	
RA	Registration Authority	
RNG	Random Number Generator	
RSA	Rivest, Shamir, Adleman	
SA	System and Services Acquisition	
SDLC	System Development Life Cycle	
SHA	Secure Hash Algorithm	
SHS	Secure Hash Standard	
SI	System and Information Integrity	
SoC	Secretary of Commerce	
SP	Special Publication	
SUT	System Under Test	
TDEA	Triple DEA	
TLS	Transport Layer Security	
URL	Uniform Resource Locator	
U.S.	United States	
U.S.C.	United States Code	
VPL	Validated Products List	
WWW	World Wide Web	

APPENDIX B

TERMS AND DEFINITIONS

This section includes terms and definitions that are used in this document. In general, the definitions are drawn from FIPS, NIST SPs, and related documents. The source of each definition is included with the definition and the full references are included in Appendix C. The source is listed at the end of the definition in square brackets [ ].

Approved: FIPS-Approved and/or NIST-recommended. An algorithm or technique that is either 1) specified in a FIPS or NIST recommendation, or 2) specified elsewhere and adopted by reference in a FIPS or NIST Recommendation. [SP 800-57]

asymmetric key algorithm: See public-key algorithm.

authentication: A process that establishes the origin of information or determines an entity's identity. [SP 800-57]

availability: Timely, reliable access to information by authorized entities. [SP 800-57]

binding: An acknowledgment by a trusted third party that associates an entity's identity with its public key. This may take place through (1) a certification authority's generation of a public key certificate, (2) a security officer's verification of an entity's credentials and placement of the entity's public key and identifier in a secure database, or (3) an analogous method. [FIPS 196]

certificate (or public key certificate): A set of data that uniquely identifies an entity, contains the entity's public key and possibly other information, and is digitally signed by a trusted party, thereby binding the public key to the entity. Additional information in the certificate could specify how the key is used and its cryptoperiod. [SP 800-57]

certificate revocation list (CRL): A list of revoked but unexpired certificates issued by a CA. [SP 800-15]

certification authority (CA): The entity in a public key infrastructure (PKI) that is responsible for issuing certificates and exacting compliance to a PKI policy. [SP 800-57]

ciphertext: Data in its encrypted form. [SP 800-57]

compromise: The unauthorized disclosure, modification, substitution or use of sensitive data (e.g., keying material and other security-related information). [SP 800-57]

confidentiality: The property that sensitive information is not disclosed to unauthorized entities. [SP 800-57]

countermeasure: An action, device, procedure, technique, or other measure.

critical security parameters: Security-related information (e.g., secret and private cryptographic keys, and authentication data such as passwords and PINs) whose disclosure or modification can compromise the security of a cryptographic module. [FIPS 140-2]

cryptographic algorithm: A well-defined computational procedure that takes variable inputs, including a cryptographic key, and produces an output. [SP 800-57]

cryptographic hash function: A function that maps a bit string of arbitrary length to a fixed length bit string. Approved hash functions satisfy the following properties:

1. (One-way) It is computationally infeasible to find any input which maps to any pre-specified output, and

2. (Collision resistant) It is computationally infeasible to find any two distinct inputs that map to the same output. [SP 800-57]

cryptographic key: A parameter used in conjunction with a cryptographic algorithm that determines its operation in such a way that an entity with knowledge of the key can reproduce or reverse the operation, while an entity without knowledge of the key cannot. Examples include:

1. the transformation of plaintext data into ciphertext data,

2. the transformation of ciphertext data into plaintext data,

3. the computation of a digital signature from data,

4. the verification of a digital signature,

5. the computation of an authentication code from data,

6. the verification of an authentication code from data and a received authentication code, and

7. the computation of a shared secret that is used to derive keying material. [SP 800-57]

cryptographic module: The set of hardware, software and/or firmware that implements Approved security functions (including cryptographic algorithms and key generation) and is contained within the cryptographic boundary. [FIPS 140-2]

cryptography: The discipline that embodies principles, means and methods for providing information security, including confidentiality, data integrity, non-repudiation, and authenticity.

cryptoperiod: The time span during which a specific key is authorized for use or in which the keys for a given system may remain in effect. [SP 800-57]

data integrity: A property whereby data has not been altered in an unauthorized manner since it was created, transmitted or stored. [SP 800-57]

DEA: The symmetric encryption algorithm that serves as the cryptographic engine for the Triple Data Encryption Algorithm (TDEA). [NIST SP 800-67]

decryption: The process of changing ciphertext into plaintext using a cryptographic algorithm and key. [SP 800-57]

DES: The symmetric encryption algorithm that serves as the cryptographic engine for the Triple Data Encryption Algorithm (TDEA). [NIST SP 800-67]

digital signature: The result of a cryptographic transformation of data which, when properly implemented, provides the services of:

1. origin authentication,

2. data integrity, and

3. signer non-repudiation. [SP 800-57]

Digital Signature Algorithm (DSA): The DSA is used by a *signatory* to generate a digital signature on data and by a *verifier* to verify the authenticity of the signature. [FIPS 186-3]

Elliptic Curve Digital Signature Algorithm (ECDSA): A digital signature algorithm that is an analog of DSA using elliptic curve mathematics and specified in ANSI standard X9.62. [SP 800-15]

encrypted key: A cryptographic key that has been encrypted using an Approved security function with a key encrypting key in order to disguise the value of the underlying plaintext key. [SP 800-57]

encryption: The process of changing plaintext into ciphertext for the purpose of security or privacy. [NIST SP 800-57]

entity: An individual (person), organization, device or process. [SP 800-57]

error detection code: A code computed from data and comprised of redundant bits of information designed to detect, but not correct, unintentional changes in the data. [FIPS 140-2]

hash function: See cryptographic hash function.

hash value: The result of applying a hash function to information. [SP 800-57]

initialization vector (IV): A vector used in defining the starting point of a cryptographic process. [SP 800-57]

integrity: The property that protected data has not been modified or deleted in an unauthorized and undetected manner. [FIPS 140-2]

key: See cryptographic key.

key encrypting key: A cryptographic key that is used for the encryption or decryption of other keys. [FIPS 140-2]

key establishment: A function in the life cycle of keying material; the process by which cryptographic keys are securely established among cryptographic modules using manual transport methods (e.g., key loaders), automated methods (e.g., key transport and/or key agreement protocols), or a combination of automated and manual methods (consists of key transport plus key agreement). [SP 800-57]

key management: The activities involving the handling of cryptographic keys and other related security parameters (e.g., IVs, counters) during the entire life cycle of the keys, including the generation, storage, establishment, entry and output, and destruction. [SP 800-57]

key pair: A public key and its corresponding private key; a key pair is used with a public key algorithm. [SP 800-57]

keying material: The data (e.g., keys and IVs) necessary to establish and maintain cryptographic keying relationships. [NIST SP 800-57]

key wrapping key: A symmetric key encrypting key. [SP 800-57]

message authentication code (MAC): A cryptographic checksum on data that uses a symmetric key to detect both accidental and intentional modifications of data. [SP 800-57]

message digest: See hash value.

non-repudiation: A service that is used to provide assurance of the integrity and origin of data in such a way that the integrity and origin can be verified by a third party as having originated from a specific entity in possession of the private key of the claimed signatory. [SP 800-57]

plaintext: Intelligible data that has meaning and can be understood without the application of decryption. [NIST SP 800-57]

private key: A cryptographic key, used with a public key cryptographic algorithm, that is uniquely associated with an entity and is not made public. In an asymmetric (public) key cryptosystem, the private key is associated with a public key. Depending on the algorithm, the private key may be used to:

1. Compute the corresponding public key,
2. Compute a digital signature that may be verified by the corresponding public key,
3. Decrypt data that was encrypted by the corresponding public key, or
4. Compute a piece of common shared data, together with other information. [SP 800-57]

public key: A cryptographic key used with a public key cryptographic algorithm, that is uniquely associated with an entity and that may be made public. In an asymmetric (public) key cryptosystem, the public key is associated with a private key. The public key may be known by anyone and, depending on the algorithm, may be used to:

1. Verify a digital signature that is signed by the corresponding private key,
2. Encrypt data that can be decrypted by the corresponding private key,
3. Compute a piece of common shared data. [SP 800-57]

public key (asymmetric) cryptographic algorithm: A cryptographic algorithm that uses two related keys, a public key and a private key. The two keys have the property that determining the private key from the public key is computationally infeasible. [SP 800-57]

public key infrastructure (PKI): A framework that is established to issue, maintain and revoke public key certificates. [SP 800-57]

RSA: One public-key algorithm used for key establishment and the generation and verification of digital signatures.

secret key: A cryptographic key that is used with a secret key (symmetric) cryptographic algorithm and is not made public. The use of the term "secret" in this context does not imply a classification level, but rather implies the need to protect the key from disclosure. [SP 800-57]

secret key (symmetric) cryptographic algorithm: A cryptographic algorithm that uses a single, secret key for an operation and its complement. [SP 800-57]

signature generation: Uses a digital signature algorithm and a private key to generate a digital signature on data. [SP 800-57]

signature verification: Uses a digital signature and a public key to verify a digital signature. [SP 800-57]

symmetric key: A single cryptographic key that is used with a secret (symmetric) key algorithm. [SP 800-57]

symmetric (secret key) algorithm: A cryptographic algorithm that uses the same secret key for an operation and its complement (e.g., encryption and decryption). [SP 800-57]

threat: An entity or event with the potential to harm a system. [NIST SP 800-12]

trusted path: A means by which an operator and a security function can communicate with the necessary confidence to support the security policy associated with the security function. [adapted from FIPS 140-2]

vulnerability: Weakness in an information system, system security procedures, internal controls or implementation that could be exploited or triggered by a threat source. [SP 800-53]

zeroization/zeroisation: A method of erasing electronically stored data by altering the contents of the data storage so as to prevent the recovery of the data. [FIPS 140-2]

APPENDIX C
REFERENCE LIST

Burr, William E., *Public Key Infrastructure (PKI) Version 1 Technical Specifications - Part C: Concept of Operations*, Federal PKI Technical Working Group, Nov. 16, 1995.

Common Criteria for Information Technology Security Evaluation, Version 2.2, International Standard ISO/IEC 15408 *Evaluation Criteria for Information Technology Security*, ISO/IEC JTC1 and Common Criteria Implementation Board.

Defense Authorization Act, Section X, Subtitle G, "Government Information Security Reform," (Public Law 106-398, Title 44 U.S. Code, Chapter 35, Subchapter II – Information Security), November 2000.

Department of Defense, *Department of Defense Trusted Computer System Evaluation Criteria*, DOD 5200.28-STD, December 1985.

Diffie, W. and M.E. Hellman, "New Directions in Cryptography," *IEEE Transactions on Information Theory*, v. IT-22, n. 6, Nov. 1976, pp. 644-654.

Dodson, D., Keller, S. S., Chang, S., and Smid. M. E., *Technical Component for the CEFMS Electronic Signature System RFP*, NIST September 22, 1992.

Electronic Signatures in Global and National Commerce Act, (Public Law 106-229), June 30, 2000.

Executive Office of the President, *Critical Infrastructure Protection*, Presidential Decision Directive 63, May 1998.

Executive Office of the President, *Critical Infrastructure Protection in the Information Age*, Executive Order, October 16, 2001.

Foti, J., Keller, S., and Dodson, Donna, *Security Review of the CEFMS Electronic Signature System*, NIST, May 17, 1996.

General requirements for the competence of testing and calibration laboratories, International Standard ISO/IEC 17025:2005, International Standards Organization, May 15, 2005.

Government Paperwork Elimination Act (GPEA), Title XVII of Public Law 105-277, October 21, 1998.

Menezes, Alfred J., vanOorschot, Paul C, and Vanstone, Scott A., *Handbook of Applied Cryptography*, CRC Press, Inc., New York, 1997.

MITRE Corporation, *Public Key Infrastructure Study, Final Report,* National Institute of Standards and Technology, April 1994.

Myers, M, R. Ankney, A. Malpani, S. Galperin, and C. Adams, *Internet Public Key Infrastructure Online Certificate Status Protocol - OSCP*, draft, Sept. 1998.

National Institute of Standards and Technology, *Advanced Authentication Technology,* NIST ITL Bulletin, 1991-12.

National Institute of Standards and Technology, *Advanced Encryption Standard*, Federal Information Processing Standards Publication 197, November 26, 2001.

National Institute of Standards and Technology, *Advanced Encryption Standard Algorithm Validation Suite (AESAVS)*, November 15, 2002.

National Institute of Standards and Technology, *AES Key Wrap Specification*, November 16, 2001.

National Institute of Standards and Technology, *AES Known Answer Test (KAT) Vectors*, April 1, 2002.

National Institute of Standards and Technology, *An Introduction to Computer Security: The NIST Handbook*, NIST SP 800-12, February 6, 1996.

National Institute of Standards and Technology, *Approved Key Establishment Techniques for FIPS 140-2, Security Requirements for Cryptographic Modules*, Federal Information Processing Standards Publication 140-2 Annex D, February 2004.

National Institute of Standards and Technology, *Approved Protection Profiles for FIPS 140-2, Security Requirements for Cryptographic Modules*, Federal Information Processing Standards Publication 140-2 Annex B, July 2003.

National Institute of Standards and Technology, *Approved Random Number Generators for FIPS 140-2, Security Requirements for Cryptographic Modules*, Federal Information Processing Standards Publication 140-2 Annex C, March 2003.

National Institute of Standards and Technology, *Approved Security Functions for FIPS 140-2, Security Requirements for Cryptographic Modules*, Federal Information Processing Standards Publication 140-2 Annex A, March 2004.

National Institute of Standards and Technology, *Computer Security Publications*, NIST Publication List 91, Revised October 1999.

National Institute of Standards and Technology, *Cryptographic Standards and Supporting Infrastructures: A Status Report*, NIST ITL Bulletin, 1997-09.

National Institute of Standards and Technology, *Digital Signature Standard (DSS)*, Federal Information Processing Standards Publication 186-3, Draft.

National Institute of Standards and Technology, *Digital Signature Standard Validation System (DSSVS) User's Guide*, June 20, 1997.

National Institute of Standards and Technology, *Entity Authentication Using Public Key Cryptography*, Federal Information Processing Standards Publication 196, February 18, 1997.

National Institute of Standards and Technology, *Federal Agency Use of Public Key Technology for Digital Signatures and Authentication,* NIST SP 800-25, October 2000.

National Institute of Standards and Technology, Federal S/MIME V3 Client Profile, NIST SP 800-49, November 2002.

National Institute of Standards and Technology, *A Framework for Cryptographic Standards,* NIST ITL Bulletin, 1995-08.

National Institute of Standards and Technology, *Generally Accepted Principles and Practices for Securing Information Technology Systems*, NIST SP 800-14, September 20, 1995.

National Institute of Standards and Technology, *Recommended Security Controls for Federal Information Systems*, NIST SP 800-53, February 2005.

National Institute of Standards and Technology, *Guide for Mapping Types of Information and Information Systems to Security Categories*, NIST SP 800-60, June, 2004.

National Institute of Standards and Technology, *Guide for the Security Certification and Accreditation of Federal Information Systems*, NIST SP 800-37, May 2004.

National Institute of Standards and Technology, *Guideline for Electronic Mail Security*, NIST SP 800-45, September 2002.

National Institute of Standards and Technology, *Guideline for the Use of Advanced Authentication Technology Alternatives*, Federal Information Processing Standards Publication 190, September 28, 1994.

National Institute of Standards and Technology, *Implementation Issues for Cryptography*, NIST ITL Bulletin, 1996-08.

National Institute of Standards and Technology, *Introduction to Public Key Technology and the Federal PKI Infrastructure*, February 2001.

National Institute of Standards and Technology, *Recommendation for Key Management*, NIST SP 800-57, August 2005.

National Institute of Standards and Technology, *The Keyed-Hash Message Authentication Code (HMAC)*, Federal Information Processing Standards Publication 198, 6 March 2002.

National Institute of Standards and Technology, *Message Authentication Code (MAV) Validation System: Requirements and Procedures,* NIST SP 500-156, May 1988.

National Institute of Standards and Technology, *Minimum Interoperability Specification for PKI Components (MISPC), NIST SP 800-15*, January 1998.

National Institute of Standards and Technology, *Modes of Operation Validation System: Requirements and Procedures*, NIST SP 800-17, February 1998.

National Institute of Standards and Technology, *Modes of Operation Validation System for the Triple Data Encryption Algorithm (TMOVS): Requirements and Procedures*, NIST SP 800-20, April 2000.

National Institute of Standards and Technology, *Public Key Infrastructure Technology*, NIST ITL Bulletin, 1997-07.

National Institute of Standards and Technology, *Recommendation for Block Cipher Modes of Operation – Methods and Techniques*, NIST SP 800-38A, December 2001.

National Institute of Standards and Technology, *Draft Recommendation for Block Cipher Modes of Operation: The CMAC Authentication Mode*, NIST SP 800-38B, May 2005.

National Institute of Standards and Technology, *Draft Recommendation for Block Cipher Modes of Operation: The CCM Model for Authentication and Confidentiality*, NIST SP 800-38C, May 2004.

National Institute of Standards and Technology, *Draft Recommendation on Key Establishment Schemes*, NIST SP 800-56, 2006.

National Institute of Standards and Technology, *Risk Management Guide for Information Technology Systems*, NIST SP 800-30, July 2002.

National Institute of Standards and Technology, *Secure Hash Standard*, Federal Information Processing Standards Publication 180-2, August 2002.

National Institute of Standards and Technology, *Security Requirements for Cryptographic Modules*, Federal Information Processing Standards Publication 140-2, May 25, 2001.

National Institute of Standards and Technology, *Standards for Security Categorization of Federal Information and Information Systems*, Federal Information Processing Standard 199 (FIPS 199), February 2004.

National Institute of Standards and Technology, *Triple DES Sample Vectors*, April 4, 2000.

National Security Agency (NSA), *Security Service API: Cryptographic API Recommendation Second Edition*, NSA Cross Organization CAPI Team, July 1, 1996.

National Security Telecommunications and Information Systems Security Committee, *National Information Systems Security Glossary*, NSTISSI No. 4009, 5 June 1992.

Office of Management and Budget, *Guidance for Preparation of Security Plans for Federal Computer Systems That Contain Sensitive Information*, OMB Bulletin No. 90-08, 9 July 1990.

Office of Management and Budget, *Guidance on Implementing the Government Information Security Reform Act*, Memorandum for the Director of OMB, January 2001.

Office of Management and Budget, *Incorporating and Funding Information System Investments*, OMB Memorandum M-00-17, February 28, 2000.

Office of Management and Budget, *OMB Guidance to Federal Agencies on Data Availability and Encryption*, November 26, 2001.

Office of Management and Budget, *Security of Federal Automated Information Resources*, Appendix III to OMB Circular No. A-130, February 8, 1996.

Office of Management and Budget, *Security of Federal Automated Information Resources*, Memorandum from the Director, June 23, 1999.

RSA Laboratories, *Diffie-Hellman Key-Agreement Standard*, Technical Note Version 1.4, PKCS #3, November 1, 1993.

Schneier, Bruce, *Applied Cryptography, Second Edition*, John Wiley & Sons, Inc., New York, c. 1996.

Smith, B. H., *Secure Electronic Grants: Key Recovery Demonstration Project Phase I Accomplishments, Test Results and Remaining Tasks*, U.S. Department of Transportation, July 24, 1998.

X/Open, *X/Open Preliminary Specification: Generic Cryptographic Service API*, draft 8, April 20, 1996.

West, John P., *Electronic Certification System (ECS) Development and Enhancement*, Financial Management Service, Department of the Treasury, April 8, 1998.

West, John P. and Chris Shanefelter, *Decision Fact Sheet, Electronic Certification System (ECS)*, Financial Management Service, Department of the Treasury, March 18, 1998.

Appendix D

Information Security Laws and Regulations

This Appendix lists standards and guidelines that apply to implementation of cryptography in the Federal government:

(a) Under the Information Technology Management Reform Act of 1996 and the Federal Information Systems Management Act of 2002 (Public Law 107-347), the National Institute of Standards and Technology (NIST) is responsible for developing technical standards and guidelines for Federal information resources.

(b) The Defense Authorization Act of 2000, Subchapter II[23], Section 3534 holds the heads of Federal Agencies responsible for 1) adequately ensuring the integrity, confidentiality, authenticity, availability and non-repudiation of information supporting agency operations and assets; 2) developing and implementing information security policies, procedures, and control techniques sufficient to afford security protections commensurate with the risk and magnitude of the harm resulting from the unauthorized disclosure, disruption, modification, or destruction of information; and 3) ensuring that the agency's information security plan is practiced throughout the lifecycle of each agency system.

(c) Public Law 106-229, Electronic Signatures in Global and National Commerce Act, promotes the use of electronic contract formation, signatures, and record keeping.

(d) Presidential Decision Directive 63, *Critical Infrastructure Protection*, May 1998, explains key elements of the administration's policy in critical infrastructure protection. PDD 63 designated NIST as the lead Agency for information and communications sector liaison.

(e) Executive Order, *Critical Infrastructure Protection in the Information Age*, 16 October 2001, authorizes a program of continuous efforts to secure information systems for critical infrastructures and states a policy of protection against the disruption of the operation of information systems for critical infrastructures. Under this order, the heads of executive branch departments and agencies are responsible and accountable for providing and maintaining adequate levels of security for information systems ... for programs under their control. The order directs cost-effective security to be built into and made an integral part of government systems and states that security should enable, and not unnecessarily impede, department and agency business operations.

---

[23] Section X, Subtitle G of the Defense Authorization Act of 2000 amends Chapter 35 of Title 44, U.S. Code by inserting Subchapter II – Information Security.

(f) Appendix III to Office of Management and Budget (OMB) Circular No. A-130 - *Security of Federal Automated Information*, in part, establishes a minimum set of controls to be included in Federal automated information security programs and assigns Federal agency responsibilities for the security of automated information. The Appendix incorporates requirements of the Computer Security Act of 1987.

(g) OMB guidance to Federal Agencies on Data Availability and Encryption, dated 26 November 2001 reports the NIST announcement of the Secretary of Commerce's approval of the Advanced Encryption Standard and notes that encryption is an important tool for protecting the confidentiality of disclosure-sensitive information entrusted to an agency's care. The guidance also notes that the encryption of agency data also presents risks to the availability of information needed by the agency to reliably meet its mission. OMB specifically states that, without access to cryptographic key(s) needed to decrypt information, the agency risks losing access to its valuable information. Agencies are reminded of the need to protect the continuity of their information technology operations and agency services when implementing encryption. The OMB guidance stresses that; in particular, agencies must address information availability and assurance requirements through appropriate data recovery mechanisms such as cryptographic key recovery.

Appendix E

Applicable FIPS and Special Publications

The following FIPS and NIST Special Publications (SP) apply to implementation of cryptography in the Federal government (URLs for documents are provided where available in electronic form):

Guidance:

NIST SP 800-21　*Guideline for Implementing Cryptography in the Federal Government*

http://csrc.nist.gov/publications/nistpubs/800-21/800-21.pdf

NIST SP 800-45　*Guideline for Electronic Mail Security*

http://csrc.nist.gov/publications/nistpubs/800-45/sp800-45.pdf

Encryption:

FIPS 197　*Advanced Encryption Standard (AES)*

http://csrc.nist.gov/publications/fips/fips197/fips-197.pdf

NIST SP 800-67　*Recommendation for the Triple Data Encryption Algorithm (TDEA) Block Cipher*

http://csrc.nist.gov/publications/nistpubs/800-67/SP800-67.pdf

NIST SP 800-38A　*Recommendation for Block Cipher Modes of Operation – Methods and Techniques*

http://csrc.nist.gov/publications/nistpubs/800-38a/sp800-38a.pdf

NIST SP 800-38C　*Recommendation for Block Cipher Modes of Operation: The CCM Model for Authentication and Confidentiality*

http://csrc.nist.gov/publications/nistpubs/800-38C/SP800-38C.pdf

Secure Hashing:

FIPS 180-2　*Secure Hash Standard (SHS)*

http://csrc.nist.gov/publications/fips/fips180-2/fips180-2withchangenotice.pdf

Digital Signatures:

FIPS 186　*Digital Signature Standard (DSS)*

http://csrc.nist.gov/publications/fips/fips186-2/fips186-2-change1.pdf

NIST SP 800-25　*Federal Agency Use of Public Key Technology for Digital Signatures and Authentication*

http://csrc.nist.gov/publications/nistpubs/800-25/sp800-25.pdf

Entity Authentication:

FIPS 196 — *Entity Authentication Using Public Key Cryptography*

http://csrc.nist.gov/publications/fips/fips196/fips196.pdf

Message Authentication:

FIPS 198 — *The Keyed HASH Message Authentication Code (HMAC)*

http://csrc.nist.gov/publications/fips/fips198/fips-198a.pdf

NIST SP 800-38B — *Recommendation for Block Cipher Modes of Operation: The CMAC Mode for Authentication*

http://csrc.nist.gov/publications/nistpubs/800-38B/SP_800-38B.pdf

NIST SP 800-38C — *Recommendation for Block Cipher Modes of Operation: The CCM Model for Authentication and Confidentiality*

http://csrc.nist.gov/publications/nistpubs/800-38C/SP800-38C.pdf

Key Management:

NIST SP 800-57 — *Recommendation for Key Management, Part 1: General Guideline and Part 2: Best Practices for Key Management Organization:*

http://csrc.nist.gov/publications/nistpubs/index.html

NIST SP 800-56 (Draft) — *Recommendation on Key Establishment Schemes*

http://csrc.nist.gov/CryptoToolkit/kms/SP800-56_7-5-05.pdf

[No Std ID] — *AES Key Wrap Specification*

http://csrc.nist.gov/CryptoToolkit/kms/AES_key_wrap.pdf

NIST SP 800-32 — *Introduction to Public Key Technology and the Federal PKI Infrastructure*

http://csrc.nist.gov/publications/nistpubs/800-32/sp800-32.pdf

Cryptographic Module Validation:

FIPS 140-2 — *Security Requirements for Cryptographic Modules*

http://csrc.nist.gov/publications/fips/fips140-2/fips1402.pdf

FIPS 140-2 Annex A — *Approved Security Functions for FIPS 140-2, Security Requirements for Cryptographic Modules*

http://csrc.nist.gov/publications/fips/fips140-2/fips1402annexa.pdf

FIPS 140-2 Annex B	*Approved Protection Profiles for FIPS 140-2, Security Requirements for Cryptographic Modules*
	http://csrc.nist.gov/publications/fips/fips140-2/fips1402annexb.pdf
FIPS 140-2 Annex C	*Approved Random Number Generators for FIPS 140-2, Security Requirements for Cryptographic Modules*
	http://csrc.nist.gov/publications/fips/fips140-2/fips1402annexc.pdf
FIPS 140-2 Annex D	*Approved Key Establishment Techniques for FIPS 140-2, Security Requirements for Cryptographic Modules*
	http://csrc.nist.gov/publications/fips/fips140-2/fips1402annexd.pdf
[No Std ID]	*Advanced Encryption Standard Algorithm Validation Suite*
	http://csrc.nist.gov/cryptval/
[No Std ID]	*AES Known Answer Test (KAT) Vectors*
	http://csrc.nist.gov/cryptval/
NIST SP 800-20	*Modes of Operation Validation System for the Triple Data Encryption Algorithm (TMOVS): Requirements and Procedures*
	http://csrc.nist.gov/publications/nistpubs/800-20/800-20.pdf
[No Std ID]	*Triple-DES Sample Vectors*
	http://csrc.nist.gov/cryptval/
[No Std ID]	*Multi-block Message Test (MMT)*
	http://csrc.nist.gov/cryptval/
NIST SP 800-17	*Modes of Operation Validation System (MOVS): Requirements and Procedures;*
	http://csrc.nist.gov/publications/nistpubs/800-17/800-17.pdf

www.ingramcontent.com/pod-product-compliance
Lightning Source LLC
Chambersburg PA
CBHW081829170526
45167CB00007B/2765